Energy Efficiency of Vehicles

Energy Efficiency of Vehicles

BY DOUGLAS R. CARROLL

Warrendale, Pennsylvania, USA

400 Commonwealth Drive
Warrendale, PA 15096-0001 USA
E-mail: CustomerService@sae.org
Phone: 877-606-7323 (inside USA and Canada)
          724-776-4970 (outside USA)
FAX   724-776-0790

Library of Congress Catalog Number 2020942782
http://dx.doi.org/10.4271/9781468601497

ISBN-Print 978-1-4686-0148-0
ISBN-PDF 978-1-4686-0149-7

To purchase bulk quantities, please contact: SAE Customer Service

E-mail: CustomerService@sae.org
Phone: 877-606-7323 (inside USA and Canada)
          724-776-4970 (outside USA)
Fax: 724-776-0790

Visit the SAE International Bookstore at books.sae.org

Chief Product Officer
Frank Menchaca

Publisher
Sherry Dickinson Nigam

Director of Content Management
Kelli Zilko

Production and Manufacturing Associate
Erin Mendicino

dedication

To my wife Karla and my children Zachary and James.

contents

CHAPTER 8

Greenhouse Gasses and How They Impact the Temperature of the Earth   173

CHAPTER 9

Fundamentals of Batteries   185

I developed a new course for the mechanical engineering program at Missouri S&T titled "Energy Efficiency of Vehicles" and offered it for the first time in the spring semester of 2017. After teaching the course three times, I wrote a manuscript for students to use, and the manuscript was converted into this book published in fall 2020.

The purpose of writing the book is to educate students and others about energy and the environment and the relationship between the energy we use and the environment. We are at a point in time when we need to make some very important decisions about energy in the next few decades, and I hope we can utilize our scientific knowledge to make good rational decisions. I do not claim to know what the right decisions are and hence, tried to make the book as nonpolitical as possible. My goal is to show how to do the calculations related to energy, power, and efficiency, and the impact of using different types of energy on the environment. None of us are willing to give up our modern lifestyle and go back to a more primitive one. Maintaining our lifestyle requires that we continue consuming large quantities of energy. The decisions that will need to be made in the coming decades are about how we will provide the energy needed for our modern lifestyle without damaging the environment.

Unit 1 of the course covers how to estimate the energy efficiency and fuel economy of many types of vehicles including bicycles, cars, trucks, trains, ships, and aircraft. In order to understand the bigger picture of energy and the environment it is necessary to have a clear understanding of the relationship between energy and power. Engineers and scientists who have been working in the energy field for several years already have a clear understanding of energy and power and will probably want to skim the first part of Unit 1. Engineering students need to build a better foundation before they can appreciate the rest of the book. Studying vehicles is a good way to develop an understanding of energy and power and how the two quantities are important in calculating the energy efficiency of vehicles. The general public tends to view energy efficiency and consumption as magic. Myths, stories, and conspiracy theories abound about cars getting 100+ mpg. I have heard such stories throughout my entire career and there are still many people who believe in them. Unfortunately, the stories are not true. Energy efficiency and consumption are not magic. Vehicles require power and energy and the amounts are related to the size of the vehicle and how fast we want to travel. There are standard methods to estimate the efficiency and fuel economy of vehicles, which are taught in Unit 1.

Unit 2 is about the environmental consequences of consuming energy. The unit starts by showing the amount of energy used in the USA and the world. This is the energy required to power our modern lifestyle. We use many different sources of energy including solar, nuclear, hydro, wind, geothermal, biomass, coal, natural gas, and petroleum. There is data that shows how much energy we use from each source. Data is also available for the carbon dioxide produced in the USA and the world and for the pollution generated while burning fossil fuels (carbon monoxide, nitrous oxide, sulfur dioxide, and particulate). It is straightforward to calculate the amount of pollution and carbon dioxide generated while burning of fossil fuels. The book compares coal, natural gas, gasoline, and diesel fuel by calculating the amount of energy we get from the fuel and the amount of pollution and carbon dioxide generated by burning the fuel. The calculations will show clearly that some fuels are cleaner than others and that there are trade-offs in choosing which fuel to use. Most of our electricity is generated using fossil fuels, and we can calculate how much pollution and carbon dioxide are generated for each kW-h of electric energy. Electric cars are sometimes said to be zero-emission vehicles, but, in this book, we will account for the pollution and carbon dioxide used in generating electricity. Wind turbines and solar panels are usually regarded as zero-emission, but manufacturing and maintaining wind turbines and solar panels impacts the environment. In Unit 2, the book covers how to calculate the environmental impact for different types of energy sources and the vehicles or other equipment that make use of the energy sources.

There is a short unit on batteries included in Unit 2. Batteries are very important for electric vehicles, and it is important to understand the basics of how batteries work. Batteries for electric vehicles need to be small and lightweight. As we move in the direction of using more wind and solar power to generate electricity, we will need to develop battery systems to help us in balancing the power grid. Utility-scale batteries are still in their infancy. Weight and size are not as critical for utility-scale batteries. Utilities will need large, reliable, inexpensive batteries. We have made a lot of progress in developing batteries for vehicles and utilities but still have a long way to go.

The purpose of Unit 3 is to understand quantitatively how city driving impacts the energy efficiency and fuel economy of cars and trucks. Spreadsheet models are developed for cars and trucks driving in city traffic to calculate quantitatively how the city driving impacts fuel economy. The models will show that regenerative braking will significantly improve the range of electric vehicles in city driving. The models will also show how hybrid electric vehicles can get a much better fuel economy in the city than traditional gasoline or diesel vehicles.

Thank you for purchasing this book. I hope you will gain an understanding and appreciation of our energy system and the impact it has on the environment.

Douglas R. Carroll

Unit 1

Introduction to Energy Efficiency

Unit 1 focuses on the energy consumption and power required by vehicles operating at their normal steady-state cruising speed. Vehicles spend most of their time cruising, and most of the energy consumed by vehicles happens while the vehicle is cruising at its normal operating speed. I included parameters in the book that will help you make reasonable estimates of the energy consumption and power requirements of many types of vehicles. At the end of Unit 1, you can make a good estimate of the energy consumption and power requirements of a vehicle based on its size, weight, speed, and type of vehicle.

The unit starts with human and animal-powered vehicles. Average humans can provide about 0.1 hp sustained power, limiting the performance of human-powered vehicles such as bicycles. Horses can provide about 1 hp pulling a buggy. For human and animal-powered vehicles we will measure the energy consumed in Cal and relate the energy to weight loss. The reader will be able to understand and calculate how exercise relates to weight loss.

Unit 1 covers the steady state driving of cars and trucks and how to estimate the fuel economy in miles per gallon (mpg). For electric vehicles, the fuel economy is measured in kW-h of energy used per mi traveled. Unit 1 does not account for the hills and stop-and-go traffic of city driving. Those topics are covered in Unit 3. Unit 1 also covers the steady-state fuel economy of trains, boats, ships, airplanes, jets, and rockets.

Energy and Power

1.1 Definition of Energy and Power

In my many years of teaching engineering courses, I have found that most people have trouble understanding the difference between energy and power. In the first section of the book, the reader must understand the difference, both qualitatively and quantitatively. In the basic physics class, the reader should have learned that energy is the dot product of force and distance:

$$Energy = Force * Distance \qquad (1.1)$$

The force may vary with the distance traveled, so it is more precise to integrate the force along the distance traveled. The dot product is necessary because the force may not be in the same direction that the vehicle is traveling.

All vehicles have a drag force associated with their motion that opposes the motion of the vehicle. For most vehicles, aerodynamic drag is an important part of the total drag force on the vehicle. Rolling vehicles have a drag associated with the tires rolling. Pedestrians experience a drag force related to the kinematics of the walking motion. Ships experience a drag force that comes from the motion of the water around the ships and the wake formed as the ship moves through the water. Airplanes and jets experience aerodynamic drag on the wings and fuselage and induced drag that is proportional to the weight of the vehicle. Some vehicles are dragged across the ground and experience a frictional drag force. The total drag force on a vehicle will be the sum of all drag forces. To move the vehicle at a steady-state, there must be a thrust force equal to the drag force. The energy required to move a vehicle through a distance is equal to the total drag force multiplied by the distance traveled.

The total drag force for most vehicles will vary along the distance traveled, especially for wheeled vehicles that will travel up and down hills and accelerate and decelerate. If the force varies with the distance traveled, Eq. 1.1 can be rewritten as an integral equation, integrating the force along the distance traveled. An equivalent approach to integration is to find the average drag force on the vehicle and multiply by the distance traveled. The average drag force for wheeled vehicles is approximately equal to the drag force at a steady speed on level ground. The uphill and downhill loads tend to cancel if the starting and ending points are at the same altitude. Acceleration and deceleration also tend to cancel if the vehicle does not do a lot of braking.

The assumption for Unit 1: Assuming that the vehicle travels at a fairly constant speed, and the altitude does not change significantly over the distance traveled, then the average drag force on the vehicle is equal to the drag force at a steady speed on level ground or level flight for aircraft. This approximation allows us to make a good estimate of the energy consumption and energy efficiency of vehicles when operating at their normal cruising speed.

Example 1.1: If the total drag force on a bicycle is 2 lb, and we want to ride the bicycle 3 mi (5280 ft/mi), how much energy is required?

$$(2\,\text{lb}) * (3\,\text{mi}) * \left(5280 \frac{\text{ft}}{\text{mi}} \right) = 31{,}680\,\text{ft-lb} \qquad (1.2)$$

To ride the bicycle 3 mi, the person riding would need to provide 31,680 ft-lb of energy. The energy comes from the carbohydrate fuel that the person consumes. It is not magic. If we can make a good estimate of the total drag force on the bicycle, we can compute how much energy is required to ride a distance, and we can calculate the number of Cal the person must consume to provide the energy. It is all related.

In the metric system, we would use newton (N) for the force unit and meter (m) for the distance. Energy would be calculated in N-m. The energy unit Joule (J) is defined as one N-m. Electric energy is usually measured in either W-h or kW-h. Human and animal energy is usually measured in nutritional Cal. (A nutritional Cal is 1000 of the calorie unit used in basic science classes. To distinguish between the two units, it is customary to use a capitol C for nutritional Cal and a lower-case c for scientific calories.) The energy in fuels is usually measured in British thermal units (BTU). Energy is such an important quantity that many different units are being used. It is always possible to convert energy from one unit to another. The table below will help make unit conversions (**Table 1.1**).

TABLE 1.1 Unit conversions
Force: 1 Pound (lb) = 4.448 Newton (N)
Distance: 1 Foot (ft) = 0.3048 meter (m)
Distance: 1 Mile = 5280 ft = 1609 m = 1.609 kilometers (km)
Energy: 1 ft-lb = 1.356 N-m = 1.356 J
Energy: 1 KW-H = 3,600,000 J = 2,655,000 ft-lb
Energy: 1 BTU = 1055 J = 778.2 ft-lb
Energy: 1 Cal = 1000 calorie = 4184 J = 3086 ft-lb
Power: 1 horsepower (hp) = 745.7 W = 0.7457 kW = 550 ft-lb/s

© SAE International

Power is the rate at which energy is consumed. If the person riding the bicycle above rides the 3 mi in 30 min (6 mph average speed), the bicycle vehicle consumes 31,680 ft-lb of energy in 30 min. The average power required is:

$$\text{Power} = \frac{31,680 \text{ ft-lb}}{(30 \text{ min}) * (60 \text{ s/min})} = 17.6 \text{ ft-lb/s} \qquad (1.3)$$

One horsepower is equal to 550 ft-lb/s, so the person riding the bicycle would need to provide 17.6/550 = 0.032 hp or 23.9 W to propel the bicycle along at 6 mph. Most people could provide that much power easily. Two pounds is a typical drag force for a bicycle ride at 6 mph. This leads to another fundamental equation for power:

$$\text{Power} = (\text{Force}) * (\text{Speed}) \qquad (1.4)$$

If we know the total drag force on the vehicle and the speed of the vehicle, we can calculate the power required to push the vehicle along. Energy is the product of power and time. The aerodynamic drag force on the vehicle will increase with the square of the speed [1], and for bicycles, the aerodynamic drag is very significant. Since force is multiplied by speed, the power requirement for the rider to overcome aerodynamic drag increases with the cube of speed. Because of this relationship, increasing the average speed of the bicycle a small amount requires a significant increase in the amount of power the rider must provide. The aerodynamic energy required to ride a distance is proportional to the square of velocity:

The aerodynamic energy required to ride a distance is proportional to the square of velocity. Energy can be expressed as the product of power and time or the product of force and distance:

$$\text{Energy} = (\text{Power}) * (\text{Time}) = (\text{Force}) * (\text{Distance}) \qquad (1.5)$$

For human-powered vehicles such as bicycles, rowboats, and canoes, the units of joules and ft-lb are useful because a joule or ft-lb is a small but significant amount of energy for a human to produce. W and ft-lb/s are small but significant amounts of power for a human to produce. Most adult humans can produce one horsepower (745.7 W) for 10–15 s. After 10–15 s, most humans will tire and must rest or reduce power output [2]. Horses can provide one horsepower for a few hours before they tire and must rest or reduce power output. A typical adult human can provide 50–100 W power for a few hours [3].

When we study electric and fuel-powered vehicles, the units of joules and ft-lb become such tiny quantities that we use other units. The kW-h is a common energy unit for electric vehicles (1 kW-h = 3,600,000 J). Hp-h is also a common energy unit, especially for farm tractors. Power is usually measured in either hp or kW.

Example 1.2: A bicycle rider faces a total drag force of 3.8 lb while riding at 10 mph on level ground. How much energy will be required to move the bicycle 1 mi (i.e. how much energy per mile traveled must the rider provide)? How much power will the rider need to provide?

Discussion: The drag force on the bicycle (and other wheeled vehicles) comes from rolling resistance in the tires and aerodynamic drag. The thin, high-pressure racing bike tires have lower rolling resistance than the wide low-pressure tires. At 10 mph and faster, most of the drag force on the bicycle will come from aerodynamics. Rolling resistance will be the dominant drag force at low speeds.

Solution: The energy and power requirement is calculated using Eq. 1.6 and Eq. 1.7.

$$\text{Energy} = (3.8\,\text{lb}) * (5280\,\text{ft}) = 20{,}064 \text{ ft-lb per mile traveled} \tag{1.6}$$

$$\text{Power} = (3.8\,\text{lb}) * \left(\frac{10\,\text{mi}}{\text{h}}\right) * \left(\frac{5280\,\text{ft}}{\text{mi}}\right) * \left(\frac{\text{h}}{3600\,\text{s}}\right) = 55.7 \text{ ft-lb/s} \tag{1.7}$$

Recognizing that one hp = 550 ft-lb/s = 745.7 W, the rider must provide 0.101 hp or 75.6 W of power to push the bicycle along at 10 mph. A typical human can provide 50–100 W of sustained power, therefore, it is reasonable to expect that a human could travel a significant distance at 10 mph. A person would need to work much harder than the woman showed **Figure 1.1** to travel at 10 mph. A well-conditioned adult can provide 150–200 W sustained power and could travel faster than 10 mph.

FIGURE 1.1   Riding a bicycle.

Example 1.3: James Watt defined the horsepower as 550 ft-lb/s. This means that a horse can produce 550 ft-lb of useful energy each s while working. It is also observed that a horse can pull a buggy along a dirt road at 7 mph (**Figure 1.2**).

a. If it takes one horsepower to pull the buggy along at 7 mph, estimate the total average drag force on the buggy.

$$\left(\frac{7\,\text{mi}}{\text{h}}\right)*\left(\frac{5280\,\text{ft}}{\text{mi}}\right)*\left(\frac{\text{h}}{3600\,\text{s}}\right)=10.27\,\text{ft}\,/\,\text{s} \tag{1.8}$$

The buggy travels 10.27 ft in 1 s, and the horse produces 550 ft-lb of energy in 1 s. Since energy is the drag force multiplied by the distance traveled (**Figure 1.3**):

$$550\,\text{ft-lb}=\left(\text{Drag Force}\right)*\left(10.27\,\text{ft}\right); \text{Drag Force}=53.6\,\text{lb} \tag{1.9}$$

b. if the buggy is to be pulled by a human that can produce 0.1 hp, how fast will the human be able to pull the buggy?

For this problem, let's use the idea that power is force times speed. These types of problems can be solved using the idea that energy = force * distance or that power = force * speed.

$$55\,\text{ft-lb/s}=\left(53.6\,\text{lb}\right)*\left(\text{Speed}\right); \text{Speed}=1.027\,\text{ft/s}=0.7\,\text{mph} \tag{1.10}$$

A human can pull with 53.6 lb and pull the buggy along the dirt road but will not be able to pull it very fast. The pulling speed is limited by the power that can be provided. The horse can pull the buggy faster because it can provide more power, not because it is pulling with a larger force. It is important to understand the difference between force, energy, and power.

FIGURE 1.2 Horse-drawn carriage.

© Shutterstock

FIGURE 1.3   Pulling a rickshaw.

Example 1.4: A small fishing boat is being pushed through the water using an electric trolling motor that provides 250 W of power. The propeller is 50% efficient in converting the rotational power into thrust power for the boat. The drag force on the boat will be proportional to the velocity squared.

$$\text{Drag Force} = (175\,\text{kg/m}) * V^2$$

How fast will the electric trolling motor pull the boat along in m/s and mph? If the battery for the trolling motor holds 1 KW-H of energy, how long will the trolling motor be able to produce this thrust force? How far (in miles) will it be able to pull the boat? (**Figure 1.4**)

The propeller will be significantly less than 100% efficient in converting the power provided to the shaft into thrust power when used to provide the thrust power. For this problem, we will assume that the electric motor is 100% efficient in providing the power to the shaft that drives the propeller. The propeller is 50% efficient in converting this power into thrust power that pushes the boat along. The electric motor will consume 250 W power from the battery. Because of the propeller, it will only provide 125 W of thrust power to the boat.

$$\text{Thrust Power} = (250\,\text{W}) * (0.50) = 125\,\text{W} \tag{1.11}$$

To solve for the speed of the boat we use the relationship that power equals drag force multiplied by speed.

$$125\,\text{W} = \left[(175\,\text{kg/m}) * V^2 \right] * V \tag{1.12}$$

$$V = 0.894\,\text{m/s} = 2\,\text{mph} \tag{1.13}$$

FIGURE 1.4 Boat with an electric trolling motor.

© Shutterstock

The boat will be able to travel at 2 mph when the trolling motor is using 250 W power from the battery. A typical trolling motor can provide up to 500 W power, so this motor is operating at about 50% power. The battery holds one KW-H of energy = 1000 W-hr. Using the idea that energy is power multiplied by time:

$$1000 \text{ W-h} = (250 \text{ W}) * (\text{Time}); \quad \text{Time} = 4 \text{ h} \tag{1.14}$$

The trolling motor would be able to operate for 4 h before the battery is depleted. Traveling for 4 h at 2 mph the boat would travel 8 mi.

Example 1.5: The Gossamer Albatross was pedal-powered across the English Channel in 1979 by Bryan Allen [4]. The aircraft was designed by Paul MacCready. The distance was 22.2 mi and it required 2 h and 49 min for the flight. Assume that Bryan is an exceptional athlete capable of providing 400 W average power during the flight. Assume that the propeller is 85% efficient in converting Bryan's power into thrust power. Estimate the drag force on the airplane in pounds (**Figure 1.5**).

This was a phenomenal achievement in its time. Airplanes require a lot of horsepowers and designing an airplane that a human could power long enough to travel across the English Channel was an amazing accomplishment. Dr. MacCready led the design effort to design an airplane that could fly slow enough with a low enough drag force that a human could power it. Power is drag force multiplied by speed, so it was important to keep drag force and speed to a minimum. Bryan Allen was an exceptional athlete, and his ability to provide lots of power was an important part of the success of the project.

FIGURE 1.5 Gossamer Albatross.

Reprinted from NASA

$$\text{Thrust Power} = (400\,\text{W}) * (0.85) = 340\,\text{W} = 250.8\,\text{ft-lb/s} \qquad (1.15)$$

$$\text{Average Speed} = \frac{22.2\,\text{mi}}{2 + \dfrac{49}{60}\,\text{h}} = 7.88\,\text{mph} = 11.56\,\text{ft/s} \qquad (1.16)$$

Recognizing that power is drag force multiplied by speed:

$$250.8\,\text{ft-lb/s} = (\text{Drag Force}) * \left(11.56\frac{\text{ft}}{\text{s}}\right); \ \text{Drag Force} = 21.7\,\text{lb} \qquad (1.17)$$

When the flight was started the winds were calm, but a headwind developed during the flight, and the airspeed of the vehicle was significantly higher than the 7.88 mph average land speed during the flight. The crew had planned for a 2 h flight, but the headwind significantly increased the flight time. The instruments on the aircraft recorded a maximum airspeed of 18 mph. If we were to add 5 mph to account for the headwind, the average airspeed would be 12.88 mph, and the average drag force would be 13.3 lb, which is probably closer to the actual drag force on the airplane.

Summary for Section 1.1: In the first section, we looked at some different vehicles and discussed energy and power and the difference between energy and power. The energy used by a vehicle is equal to the total drag force on the vehicle multiplied by the distance traveled. For human-powered vehicles, the energy comes from the human and will relate to the number of Cal burned in powering the vehicle the distance traveled. For vehicles powered by an internal combustion engine, the energy will relate to the amount of fuel consumed over the distance traveled. For electric vehicles, the energy will relate to the electric energy drawn out of the batteries in powering the vehicle the distance traveled.

The power required for a vehicle is equal to the product of drag force and speed. The power requirement for a vehicle increases with speed, even if the drag force remains constant. The drag force for most vehicles also increases with speed, so the power requirement of the vehicle increases significantly as speed increases.

1.2 **Basics of Thermal Efficiency [5]**

Power is the rate at which the engine (or human or animal) can provide energy. Power is also the rate that the vehicle uses energy. For steady-state operation, the engine must provide the power required for the vehicle. Engines, motors, humans, and animals are not 100% efficient in converting fuel into useful energy to power the vehicle. For internal combustion engines, most of the heat energy in the fuel goes out the tailpipe. The same is true for jet engines, rocket engines, the steam turbines used at the power plants, and all heat engines. Less than half of the energy in the fuel will be converted to useful energy. We refer to this as the thermal efficiency of the engine. If the engine converts 28% of the energy in the fuel to useful mechanical energy, we say the engine has a thermal efficiency of 28%. To estimate the fuel consumption of an engine we need to know the thermal efficiency.

Humans and animals have a thermal efficiency of about 20% [6]. About 80% of the carbohydrate fuel we consume is converted to heat, and we heat up as we exercise. Only about 20% of the carbohydrate fuel is converted into useful mechanical energy. The thermal efficiency comes from the way our muscles operate and how they use the carbohydrate fuel.

Electric motors and generators have high thermal efficiency. A typical electric motor will convert 85% of the electric energy it uses into useful mechanical energy. Only about 15% of the electric energy will convert into heat. High-efficiency electric motors can have a thermal efficiency of 95%, or perhaps even a little higher. No electric motor is 100% efficient. In the power plant, we use a steam turbine to power a generator that is highly efficient in converting the mechanical energy from the steam turbine into electrical energy. The steam turbine has lower thermal efficiency, typically 35% to 40%, in converting the energy in the fuel (coal, natural gas, nuclear, etc.) into mechanical energy. The overall efficiency of the power plant in converting the energy in the fuel into electric energy is typically about 33%. When we use electricity to power our vehicles, the electricity is produced from other energy sources. The overall energy efficiency in getting useful energy from the fuel source is not much different for electric vehicles and internal combustion vehicles.

> **Example 1.6:** When we talk about diet and Cal, a Cal is 4,184 Joules. It is common knowledge that a human at rest emits about the same heat as a 100-W light bulb. If a human is using 100-W power while at rest, how many Cal would the person consume during the day, assuming the person sat around doing nothing all day? (Cal = 1000 calorie = 4184 Joules)
>
> $$100\,\text{W} = \left(\frac{100\,\text{J}}{\text{s}}\right) * \left(\frac{3600\,\text{s}}{\text{h}}\right) * \left(\frac{24\,\text{h}}{\text{day}}\right) * \left(\frac{\text{Cal}}{4184\,\text{J}}\right) = 2065\,\text{Cal}\,/\,\text{day} \qquad (1.18)$$

People use less than 100 W while sleeping, so this may be a high estimate. For a very sedentary person, an estimate of 2000 Cal per day is reasonable. When we talk about weight and nutrition 1-lb of weight (fat) is equal to about 3500 Cal. A person would need to consume about 3500 Cal less than they burn to lose 1 lb. Or a person who consumes 3500 Cal more than they burn will gain 1-lb of weight. Physical activity will increase the number of Cal burned in a day but not as much as we would hope. The example problems and homework will help you gain an understanding of how exercise affects weight loss.

Energy is very important in all fields of science, and there are many different units used for energy. Google can be very helpful in looking up the unit conversions. Learning to convert energy units is part of working in the energy field.

Example 1.7: The aerodynamic drag force on a car is proportional to the square of its velocity. For a typical mid-sized sedan, the aerodynamic drag force F is given by:

$$F = (0.0262)V^2 \qquad (1.19)$$

Where F is in lb and V is in mph.

Find the extra energy required for a car to go on a 250-mi trip if the car travels at 80 mph rather than 70 mph. That is, find the aerodynamic energy required if the car travels at 70 mph, at 80 mph, and subtract the two to find the difference (**Figure 1.6**).

$$\text{At } 70 \text{ mph, Drag Force} = (0.0262)(70)^2 = 128.4 \text{ lb} \qquad (1.20)$$

$$\text{At } 80 \text{ mph, Drag Force} = (0.0262)(80)^2 = 167.7 \text{ lb} \qquad (1.21)$$

Equations 1.23 and 1.24 show that the aerodynamic drag force on the car is substantially higher at 80 mph than at 70 mph. Energy consumed on a trip is drag

FIGURE 1.6 Mid-sized sedan.

force multiplied by the distance traveled, and the distance, in this case, is 250 mi. The energy required to overcome aerodynamic drag is calculated as follows:

$$\text{At 70 mph, Drag Energy} = (128.4\,\text{lb})(250\,\text{mi})(5280\,\text{ft / mi}) = 169.5 \times 10^6\,\text{ft-lb} \quad (1.22)$$

$$\text{At 80 mph, Drag Energy} = (167.7\,\text{lb})(250\,\text{mi})(5280\,\text{ft / mi}) = 221.4 \times 10^6\,\text{ft-lb} \quad (1.23)$$

$$\text{Extra Energy} = 221.4 \times 10^6 - 169.5 \times 10^6 = 51.9 \times 10^6\,\text{ft-lb} \quad (1.24)$$

This is the correct answer, but we need to understand what it means. Is this a lot of energy or a small amount of energy? For humans or horses, this is a lot of energy. One horsepower is 550 ft-lb/s of power. One way to look at this is to solve for the number of hours it would take a horse to produce this much energy. Solving for the time t required:

$$(550\,\text{ft-lb/s})(t) = 51.9 \times 10^6\,\text{ft-lb}, \quad t = 94{,}364\,\text{s} = 26.2\,\text{h} \quad (1.25)$$

We would need to work a horse for 26.2 h to produce enough energy to provide the excess of energy required for the car to make the trip at 80 mph rather than 70 mph. Horses can be worked 4–5 h per day, so it would take several days for a horse to produce this much energy. For horses, humans, or any animal, 51.9 × 10⁶ ft-lb represents a lot of energy.

For cars and trucks, we need to look at the energy content of the fuel. **Table 1.2** below shows the energy content of many types of fuel [7].

TABLE 1.2 The energy content of common fuels

Fuel Type	MJ/L	MJ/kg	BTU/Imp gal	BTU/US gal
Regular gasoline	34.8	~47	150,100	125,000
Premium Gasoline	34.8	~46	150,100	125,000
Autogas (LPG) (60% propane, 40% butane)	25.5–28.7	~51	N/A	N/A
Ethanol	23.5	31.1	101,600	84,600
Gasohol (10% ethanol, 90% gasoline)	33.7	~45	145,200	121,000
E85 (85% ethanol, 15% gasoline)	25.2	~33	108,878	90,660
Diesel	38.6	~48	166,600	138,700
Biodiesel	35.1	39.9	151,600	126,200
Aviation gasoline	33.5	46.8	144,400	120,200
Jet fuel (naphtha)	35.5	46.6	153,100	127,500
Jet fuel (kerosene)	37.6	~47	162,100	135,000
Liquefied natural gas	25.3	~55	109,000	90,800
Liquid hydrogen	9.3	~130	40,467	33,696

© SAE International

Cars and trucks purchase fuel by the gallon, so the most important column for cars and trucks in the column that shows BTU/US gal. Gasoline has 125,000 BTU/gal, but in the Midwest, most of the gasoline we purchase has a 10% ethanol content and is called gasohol in the table. Gasohol has an energy content of 121,000 BTU/gal, which is a little less than gasoline. Many gas stations carry the E85 fuel, which has an energy content of 90,660 BTU/gal. The fuel economy of a vehicle, as measured in mpg, is proportional to the energy content of the fuel. A car or truck will get slightly lower fuel economy burning gasohol compared to burning regular gasoline. The car will get significantly lower fuel economy burning the E85 fuel. Diesel fuel has an energy content of 138,700 BTU/gallon, which is significantly higher than gasoline. Part of the reason that diesel vehicles get better fuel economy than gasoline vehicles is because diesel fuel has more energy per gallon. Diesel engines have a higher thermal efficiency than gasoline engines, which also gives them a fuel economy advantage over gasoline-powered vehicles. The result is diesel-powered vehicles get significantly better fuel economy than gasoline-powered vehicles.

BTU Definition: Heating water is an important use of energy. The BTU is a unit of energy originally designed to support the application of heating water. A BTU is the amount of energy required to heat 1-lb of water 1 °F. A pint of water weighs about 1 lb. To raise the temperature of a pint of water 1 °F requires 1 BTU of heat energy.

Example 1.8: How much energy is required to heat a quart of water from 65 °F to 150 °F? Assume that the quart of water has a weight of 2 lb for this example (**Figure 1.7**).

$$\left(2\,\text{lb}\right)\left(150\,°\text{F} - 65\,°\text{F}\right)\left(\frac{1\,\text{BTU}}{\text{lb}\,°\text{F}}\right) = 170\,\text{BTU} \tag{1.26}$$

A lot of heat goes into the air and stove. The stove will need to provide more heat energy than 170 BTU to heat the water, but this type of analysis is a good place to start when estimating the heat energy required.

When estimating the energy required for vehicles we estimate in units of Joules and ft-lbs. The energy content of the fuel is usually given in BTU, so converting from BTU to Joules and ft-lb is common.

$$\text{BTU} = \left(1055\right)\text{Joule} = \left(778.2\right)\text{ft-lb} \tag{1.27}$$

FIGURE 1.7   Boiling a pot of water.

© Shutterstock

To continue building depth in the subject it is necessary to have a brief discussion of thermodynamics. The first and second laws of thermodynamics are important in understanding the energy efficiency of vehicles[5].

First Law: Energy and Energy Conversion: Energy is conserved. Conservation of energy is the first law of thermodynamics. There are many different types of energy, and it is possible to convert from one type to another. The important types of energy for this subject follow:

Chemical Energy (fuel) – especially fossil fuel: This energy can be easily converted to heat by igniting the fuel. It is not (from a practical viewpoint) possible to convert the heat back into fuel. The energy is usually measured in BTU or KW-H. 1 WH = 3.6 million Joules = 3412 BTU = 860.4 Cal = 2.655 million ft-lbs.

Electrical Energy: Electricity is produced primarily from fossil fuel at the power plant. Significant amounts of electricity are produced by nuclear, hydro, and wind. Electricity can be stored in batteries as chemical energy. Electric energy is usually measured in KW-H.

Gravitational Energy: Gravitational energy is conservative in the sense that it requires energy to push a vehicle uphill, but the energy is recovered when the vehicle goes downhill.

Kinetic Energy: Kinetic energy is also conservative. Energy is required to speed the car up and is stored as kinetic energy. On an electric vehicle, the kinetic energy can be recovered with regenerative braking, but for most vehicles, the kinetic energy is converted to heat using the brakes.

Friction Energy: A friction force absorbs energy. Friction converts mechanical energy into heat energy. It is not reversible. Braking, rolling resistance, and aerodynamic drag fall under the general category of friction energy.

Second Law of Thermodynamics and Heat Engines: The second law of thermodynamics says that, when we convert from one form of energy to another, some of the energy will be converted to waste heat and (for all practical purposes) lost. All energy conversions are less than 100% efficient.

Converting from electric energy to kinetic energy using an electric motor is very efficient. At least 85% of the electric energy will be converted into mechanical and, ultimately, kinetic energy. Converting from mechanical energy to electric energy using a generator is equally efficient.

Charging and discharging a battery is converting from electrical energy to chemical energy and back to electrical energy. This is also very efficient. Lead-acid batteries are typically 80% efficient or better in storing electric energy. Batteries will heat up when being charged and discharged, and the heat energy is the energy that is "lost" in the charge/discharge cycle.

Heat engines are inefficient in converting chemical energy in fuel into mechanical energy. The steam turbines used in power plants are about 40% efficient at converting heat energy in coal or natural gas into mechanical energy. Further, inefficiencies in the generator, transmission, and distribution make the overall process about 35% efficient in converting the energy in fuel into useful electric power at your house. The internal combustion engines in our vehicles are less efficient than the steam turbine since they must operate over a range of speeds. Under load on the highway, 25% is good efficiency for a gasoline engine, and 35% efficiency is good for a large diesel truck engine.

Example 1.9: Going back to the problem of the car traveling 250 mi at 80 mph or 70 mph, it was discovered that it would require an extra 51.9×10^6 ft lb of energy to make the trip. If it is a gasoline-powered car operating at 25% efficiency, how many extra gallons of fuel will be required for the trip? Assume the gasoline has 121,000 BTU of energy per gallon.

$$\left(\frac{121{,}000\,\text{BTU}}{\text{gal}}\right)\left(\frac{778.2\,\text{ft-lb}}{\text{BTU}}\right) = \frac{94.16 \times 10^6\,\text{ft-lb}}{\text{gal}} \left(\text{Total Energy per gallon}\right) \quad (1.28)$$

If the gasoline engine in the car had a thermal efficiency of 100%, then each gallon of the gasohol fuel would provide 94.16×10^6 ft-lb of useful energy. However, because of the laws of thermodynamics, the engine will be much less than 100% efficient. An efficiency of 25% is typical for a gasoline engine. This means that 75% of the energy in the fuel is going out the tailpipe, into the radiator, or radiated elsewhere as heat. Only 25% of the energy in the fuel is being converted to useful mechanical energy that can be used to push the car down the road.

$$\left(\frac{94.16 \times 10^6\,\text{ft-lb}}{\text{gal}}\right)(0.25) = \frac{23.54 \times 10^6\,\text{ft-lb}}{\text{gal}} \left(\text{Useful Energy per gallon}\right) \quad (1.29)$$

If we divide the required energy by the useful energy per gallon, we can find the number of gallons of fuel required.

$$\frac{51.9 \times 10^6\,\text{ft-lb}}{23.54 \times 10^6\,\text{ft-lb/gal}} = 2.205\,\text{gal} \quad (1.30)$$

The final answer is that, if we choose to drive the car at 80 mph rather than 70 mph it will require that we use an extra 2.205 gals of fuel to make the 250-mi trip. The extra cost of driving the trip at 80 mph vs 70 mph would be about 2.2 gals of fuel.

The next question to be addressed is how much total fuel the car would use in making the 250-mi trip and how to calculate the fuel economy in mpg. From Eq. 1.25 and Eq. 1.26 we found that the energy to overcome the aerodynamic drag for the car was 169.5×10^6 ft-lb at 70 mph and 221.4×10^6 ft-lb at 80 mph. The car will have rolling resistance too, and there will be additional energy required to overcome the rolling resistance which is not included in the analysis.

$$70\,\text{mph} -> \frac{\left(169.5 \times 10^6\,\text{ft-lb}\right)}{23.54 \times 10^6\,\text{ft-lb/gal}} = 7.200\,\text{gal}; \quad \frac{250\,\text{mi}}{7.200\,\text{gal}} = 34.7\,\text{mpg} \quad (1.31)$$

$$80\,\text{mph} -> \frac{\left(221.4 \times 10^6\,\text{ft-lb}\right)}{23.54 \times 10^6\,\text{ft-lb/gal}} = 9.405\,\text{gal}; \quad \frac{250\,\text{mi}}{9.405\,\text{gal}} = 26.6\,\text{mpg} \quad (1.32)$$

This analysis accounts only for the aerodynamic drag. If we add the energy required to overcome rolling drag the fuel economy of the vehicle will be lower in both cases. The next step in the analysis is to consider an electric car traveling the 250-mi trip and estimate the number of batteries required to make the trip.

Example 1.10: Going back to the problem of the car traveling 250 mi at 80 mph or 70 mph, it was discovered that it would require 169.5×10^6 ft-lb at 70 mph and 221.4×10^6 ft-lb at 80 mph to make the trip. If it is an electric car, and the motor is 85% efficient, how many lead-acid batteries would be required to make the trip at 70 mph and 80 mph?

    a. Assume that the lead-acid batteries weigh 80 lb and store 1000 W-h of energy each. These are typical values for a 12V 80 amp-h battery that might be used to power a trolling motor on a fishing boat. How many lead-acid batteries are required and how much would the battery system weigh?

    b. Suppose lithium-ion batteries are used that weigh 20 lb each and store 1000 W-h each. How many lithium-ion batteries are required and how much do they weigh? (**Figure 1.8**)

At 70 mph the energy required to make the 250-mi trip was 169.5×10^6 ft-lb. If the battery and electric motor system are 85% efficient, we would need:

$$\frac{169.5 \times 10^6 \text{ ft-lb}}{0.85} = 199.4 \times 10^6 \text{ ft-lb energy } \left(\text{Energy in Batteries}\right) \qquad (1.33)$$

To calculate how many batteries are required we need to relate the energy units of ft-lb and W-h. The conversion is 2655 ft-lb = 1 W-h.

$$\frac{199.4 \times 10^6 \text{ ft-lb}}{2655 \text{ ft-lb/W-h}} = 75,090 \text{ W-h } \left(\text{Energy in Batteries}\right) \qquad (1.34)$$

Since each battery holds 1000 W-h of energy, we would need 76 batteries fully charged to provide the required energy to allow the car to travel the 250 mi. For this

FIGURE 1.8 Lithium-ion battery.

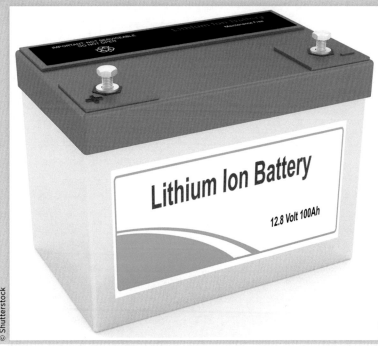

example, the lead-acid and lithium-ion batteries hold the same amount of energy, so it is 76 of either type of battery.

Lead-acid batteries weigh 80 lb each, so the total weight of the lead-acid batteries would be 6080 lb. A typical sedan weight is 3500 lb, so the batteries would be much heavier than the car. This illustrates the problem of using lead-acid batteries to power an electric car. To provide enough energy to power the car a reasonable distance, the batteries are very heavy. As the car gets heavier, the rolling resistance energy increases. It becomes impractical to make a lead-acid powered car that can travel 250 mi on a single charge.

Lithium-ion batteries weigh 20 lb each, so the battery weight required to provide the energy to travel 250 mi is 1520 lb. Although adding 1520 lb of batteries to a 3500-lb vehicle is a lot, it might be possible to make a lithium-ion battery-powered car that can travel 250 mi on a single charge.

Gasoline has a density of 6.65 lb/gal. The weight of the 7.200 gals to make the trip is 48 lb. If we do a comparison, 6080 lb of lead-acid batteries, 1520 lb of lithium-ion batteries, or 48 lb of gasoline is required to make the 250-mi trip. Lead-acid batteries are 126 times as heavy as gasoline, and lithium-ion batteries are 62.7 times as heavy as gasoline. This is the primary challenge in developing electric cars. The batteries are very heavy compared to gasoline.

Traveling at 80 mph, the energy required increases, and more batteries are required. Similar calculations show that 99 batteries are required, and the weights for lead-acid and lithium-ion batteries are 7920 lb and 1980 lb, respectively.

Example 1.11: (This example problem was developed for the Honda 750 motorcycle. The equations and results are correct for the Honda 750). Assume that a motorcycle and rider have a total weight of 700 lbs. The drag force on the motorcycle is given by the equation:

$$\text{Force (lb)} = 5.6 + (0.019) * V + (0.0108) * V^2 \quad \text{(where V is in ft/s)} \quad (1.35)$$

What is the fuel economy of the motorcycle cruising at 70 mph? Assume that the fuel is gasohol and the engine/drive system is 22% efficient in converting the chemical energy in the fuel into useful mechanical energy (**Figure 1.9**).

For this problem, we are assuming the motorcycle is driving on level pavement. The force equation accounts for rolling and aerodynamic drag and is a good estimate of the total drag force on the motorcycle. The motorcycle has an air-cooled engine that will have a lower thermal efficiency than a water-cooled engine. The estimate of 22% thermal efficiency is reasonable. First, we need to convert 70 mph into ft/s.

$$\left(\frac{70\,\text{mi}}{\text{h}}\right)\left(\frac{5280\,\text{ft}}{\text{mi}}\right)\left(\frac{\text{h}}{3600\,\text{s}}\right) = 102.7\,\text{ft/s} \quad (1.36)$$

$$\text{Force} = 5.6 + (0.019) * (102.7) + (0.0108) * (102.7)^2 = 121.4\,\text{lb} \quad (1.37)$$

FIGURE 1.9   Honda 750 motorcycle.

© Shutterstock

In the force equation, the first two terms come from the rolling drag, and the third term comes from aerodynamic drag. Going term by term the drag force on the motorcycle is given by:

$$\text{Force} = 5.6 + 1.95 + 113.84 = 121.4\,\text{lb} \tag{1.38}$$

When going fast on a motorcycle, most of the energy goes into overcoming the aerodynamic drag. In this case, of the 121.4 lb of total drag force on the motorcycle, 113.84 lb comes from the aerodynamic drag. Thus, the aerodynamic drag accounts for 93.8% of the total drag force on the motorcycle. For slower driving, the aerodynamic drag term is much less, and rolling resistance is a larger part of the total drag. The same is true for bicycles and scooters, except the definition of fast would be 15 mph for bicycles and 30 mph for scooters. When going fast on a two-wheeled vehicle, most of the energy goes in overcoming the aerodynamic drag. The power required to push the motorcycle along at highway speed is force multiplied by speed.

$$\text{Power} = (121.4\,\text{lb})(102.7\,\text{ft/s}) = 12,468\,\text{ft-lb/s} = 22.7\,\text{hp} \tag{1.39}$$

We will assume that the motorcycle is burning gasohol which has 121,000 BTU per gallon. The thermal efficiency of the engine is 22%. From these numbers we can calculate the useful energy per gallon in the fuel, the fuel consumption for the motorcycle, and the fuel economy as shown below:

$$\text{Useful Energy} = (121,000)(.22) = 26,620\,\text{BTU / gal} \tag{1.40}$$

$$\text{Fuel Consumption} = \left(\frac{12,468\,\text{ft-lb}}{\text{s}}\right)\left(\frac{\text{BTU}}{778.2\,\text{ft-lb}}\right)\left(\frac{\text{gal}}{26,620\,\text{BTU}}\right)\left(\frac{3600\,\text{s}}{\text{h}}\right) = 2.167\,\text{gal/h} \tag{1.41}$$

$$\text{Fuel Economy} = \left(\frac{70\,\text{mph}}{2.167\,\text{gph}}\right) = 32.3\,\text{mpg} \tag{1.42}$$

Notice in the final step that, if we know the speed of the vehicle in mph and the fuel consumption in gal/h (gph), the fuel economy is speed divided by fuel consumption. This approach will be used for all the fuel-powered vehicles studied in this chapter. The calculations show a fuel economy of 32.3 mpg, which is very close to what is typically achieved for a Honda 750 motorcycle. Continuing with the same motorcycle, what is the top speed of the motorcycle, assuming the engine/drive can deliver a maximum of 40 hp to the rear wheel?

$$(40\,\text{hp})(550\,\text{ft-lb/s hp}) = 22{,}000\,\text{ft-lb/s} \tag{1.43}$$

And since power is force multiplied by velocity, we can set up an equation to solve for the top speed of the motorcycle:

$$22{,}000\,\text{ft-lb/s} = \left[5.6 + (0.019) * V + (0.0108) * V^2\right] * V \tag{1.44}$$

Solving this equation, the top speed of the motorcycle is 124.8 ft/s, which is the same as 85.1 mph. Holding the engine at full throttle on level pavement, the motorcycle would top out at about 85 mph, which is typical for a Honda 750 motorcycle. We can use this approach to plot the power vs speed curve for the motorcycle which is illustrated in the graph below (**Figure 1.10**).

At higher speeds and power requirements the thermal efficiency of the engine will be near 22%, and the fuel economy analysis is reasonable. At lower speeds and power requirements, the thermal efficiency of the engine will be less than 22%, and the analysis used in this unit will yield a fuel economy estimate that is too high.

FIGURE 1.10   Power vs speed for Honda 750 motorcycle.

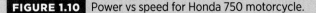

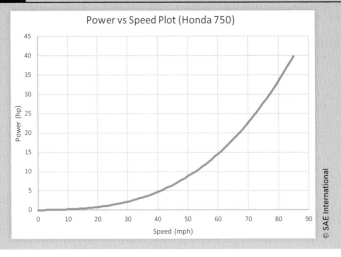

© SAE International

In Unit 1, we will focus on estimating the fuel economy of vehicles at their cruising speeds. Trains, ships, airplanes, and jets tend to get up to their cruising speed and spend most of the trip at cruising speed. Estimating the fuel economy at cruising speed is a good way to estimate the overall fuel economy of the vehicle. Cars and trucks driving on the highway spend most of their time at cruising speed. The analysis in Unit 1 is useful in estimating the fuel economy for highway driving of cars and trucks but overestimates the fuel economy in city driving. In Unit 3, we will study the details and model the city driving of cars and trucks.

1.3 **Homework**

To understand the material covered so far, the reader needs to solve some homework problems. Partial answers are given for the problems to help the reader ensure correct processes that are used in solving the problems. The first chapter in Unit 1 covers the relationships between drag force, speed, energy consumption, power, and fuel economy that apply to all types of vehicles. Subsequent chapters go into detail on how to estimate these quantities for different types of vehicles. At the end of Unit 1, the reader will be able to make good estimates of the drag force, power, energy consumption, and energy efficiency for many types of vehicles based on the size, shape, weight, and speed of the vehicle.

1. Suppose that a small ultralight is powered by a gasoline engine. The cruising speed for the ultralight is 33 mph.

   a. Assume that it requires 25 hp of thrust to power the ultralight at 33 mph and calculate the drag force on the ultralight at 33 mph. Express your answer in lb.

   b. Calculate the energy required for the ultralight to travel for 2 h (66 mi) at 33 mph. Express your answer in BTU.

   c. The gasoline has an energy density of 121,000 BTU per gallon. Assuming the engine and propeller are 20% efficient in converting the energy in the gasoline into thrust energy, how large should the fuel tank be to allow for 2 h of cruising? (Answer 5.26 gals. We would probably need 6 gals to allow for take-off and landing.) (**Figure 1.11**)

| FIGURE 1.11 | Ultralight aircraft.

© Shutterstock

2. In the Iditarod, a sled is being pulled by 10 dogs each capable of producing 0.2 hp. It is observed that the sled is traveling along at 4 mph on level ground.

   a. What is the average drag force on the sled?

   b. How fast would the sled travel if only 8 dogs were pulling? (Answer 3.2 mph) (**Figure 1.12**)

FIGURE 1.12 Sled with sled dogs.

3. A person decides to participate in a 50-mi bicycle fundraiser as a community service project. The event starts at 8:00 in the morning, and the person finishes the event at 3:30 in the afternoon. In the event, all riders are required to take a 1 h break at the halfway point to rehydrate. Moreover, the person will have several short stops along the way to stop signs and other minor reasons. Assume an additional 30 min of stopped time to account for these minor stops.

   a. Subtracting out the break time and the minor stops, what is the average speed of the person while riding?

   b. The drag force on the person comes from aerodynamics and rolling drag. With the speed V in mph, assume that the drag force in pounds is given by the following equation:

$$F = (0.016)V^2 + 1.8 \text{ lb} \tag{1.45}$$

   c. Using the average speed while traveling above, how much energy will be required for the person to travel the 50 mi? Please express your answer in ft-lbs.

d. One Cal is 3088 ft-lbs of energy. Our leg muscles are about 20% efficient in converting the sugars into useful work so, to produce the energy above, a person will need to burn about 5 times that much Cal energy. How many Cal will the person burn during the 50-mi bike ride?

e. There are approximately 3500 Cal associated with 1 lb of weight. If a person were to make this 50-mi bike ride 10 times during a month, how much weight loss would be associated with riding the bike? (Answer 3.56 lb) (**Figure 1.13**)

FIGURE 1.13   Riding bicycles.

© Shutterstock

4. It is observed that a 49cc scooter gets 90 mpg while traveling at 30 mph. If we assume the engine is 22% efficient in converting the heat energy in the fuel to the energy used to propel the scooter along, what is the average drag force on the scooter? Assume the scooter is burning 10% ethanol gasoline. (Answer: 43.6 lb) (**Figure 1.14**)

5. Driving into a gas station, you notice that the price of the regular unleaded (10% ethanol blend) is $2.29 per gallon, and the E85 gasoline (85% ethanol blend) is priced at $1.99 per gallon. On average, your vehicle has been getting 18.3 mpg using the regular unleaded (10% ethanol) gas.

a. What is the fuel cost per mile in driving your vehicle using regular unleaded gasoline?

b. Assuming your car is designed to be able to use E85 (flex fuel vehicle), what would you estimate your fuel economy to be using E85 gasoline? Based on this number, what is your fuel cost per mile using E85? (Answer: 13.7 mpg)

c. Which fuel will give you the overall lowest fuel cost per mile?

FIGURE 1.14 Scooter.

6. A cruise ship requires 45,000 hp from the engines to propel itself along at 26 mph. The propellers driving the ship are 85% efficient, so the actual power delivered to the water to provide thrust is 38,250 hp.

   a. Calculate the drag force on the cruise ship. Express your answer in pounds.

   b. If the engines are 30% efficient and the fuel has an energy density of 138,700 BTU/gallon, how many gallons of fuel are required to travel 300 mi? (Answer 31,750 gals) (**Figure 1.15**)

FIGURE 1.15 Cruise ship.

References

1. Munson et al., *Fundamentals of Fluid Mechanics*, John Wiley & Sons, 1994. Or any good fluid mechanics textbook.

2. Wikipedia has a good article on human power. https://en.wikipedia.org/wiki/Human_power

3. Wikipedia article on horsepower. https://en.wikipedia.org/wiki/Horsepower.

4. Wikipedia article about the Gossamer Albatross. https://en.wikipedia.org/wiki/MacCready_Gossamer_Albatross

5. Moran et al., *Fundamentals of Engineering Thermodynamics*, Wiley, 2014. Or any good thermodynamics textbook.

6. http://mb-soft.com/public2/humaneff.html

7. https://en.wikipedia.org/wiki/Energy_density Selected information from the source.

CHAPTER 2

Wheeled Vehicles

2.1 Thermal Efficiency of Internal Combustion Engines

The first term that needs to be defined is the thermal efficiency of internal combustion engines. The thermal efficiency of the engine is the percentage of the chemical energy in the fuel that is converted into useful work. Most of the chemical energy in the fuel will be converted to heat. A thermal efficiency of 25% is good for gasoline engines for highway driving. Car companies have made gradual improvements in the thermal efficiency of gasoline engines, and the newest cars have engines that approach 30% thermal efficiency for highway driving. The diesel engines used in large trucks can approach a thermal efficiency of 35% under highway conditions and are among the most efficient internal combustion engines in service.

The thermal efficiency of an internal combustion engine is highest when the throttle is open and a lot of power is being generated by the engine. The thermal efficiency of the engine is zero at idle. Thermal efficiency is lower for city driving and higher for highway driving.

Fuel economy is related to thermal efficiency but is not the same as thermal efficiency. Fuel economy is normally measured in miles per gallon (mpg). Most newer cars have an instantaneous fuel economy meter on the dash for the driver to view. Since the computer in the fuel injection system knows the rate that the car is using fuel (gallons per hour), and since the car speed is also known by the computer system (miles per hour), the instantaneous fuel economy is calculated as the car speed divided by the fuel consumption.

The instantaneous fuel economy gauge can be misleading. When accelerating, the thermal efficiency of the engine is good but the fuel economy is poor. We push on the gas pedal (open the throttle) to accelerate the vehicle and this increases the gph fuel consumption. Since fuel economy is speed divided by fuel consumption, the fuel economy during acceleration is low. Opening the throttle improves the thermal efficiency of the engine, so the thermal efficiency of the engine when accelerating is good.

When coasting, the thermal efficiency of the engine is poor but the fuel economy is good. When we let off the gas pedal (close the throttle), we reduce the gph fuel consumption of the engine. Since fuel economy is speed divided by fuel consumption, the fuel economy during coasting or braking is high. Closing the throttle reduces the thermal efficiency of the engine, so the thermal efficiency of the engine when coasting or braking is low. Higher thermal efficiency will correspond to better fuel economy but the instantaneous fuel economy readings are misleading.

As the car speeds up, the drag force increases and the engine must produce more power. Higher power output from the engine increases the thermal efficiency, and higher thermal efficiency tends to increase the fuel economy. As the car speeds up, the drag force increases and tends to reduce the fuel economy of the vehicle. Achieving the best fuel economy requires that we find a balance between good thermal efficiency and low drag force.

At lower speeds, the improvement in thermal efficiency is more important than the increase in the drag force. For a typical car or truck, driving at 30 mph will yield better fuel economy than driving at 20 mph even though the drag force is higher at 30 mph. The engine has higher thermal efficiency at 30 mph than at 20 mph.

Most cars and trucks driving at 80 mph will yield a lower fuel economy than driving at 70 mph. The engine has higher thermal efficiency at 80 mph, but the increase in drag force at 80 mph more than offsets the improvement in thermal efficiency. Typical cars and trucks get the best fuel economy at 55–60 mph.

Electric motors have zero efficiencies at zero speed, but their efficiency increases rapidly with speed and then levels out at about 15–20 mph. Electric cars will get their maximum economy (KW-H/mile) at much slower speeds than internal combustion engine-powered cars.

2.2 Rolling and Aerodynamic Drag for Wheeled Vehicles

Power is the rate at which a vehicle uses energy. The common units for power are hp, ft-lb/s, W, and kW. For unit conversions, hp = 550 ft-lb/s = 745.7 W = 0.7457 kW. For internal combustion engines, the power is normally given in hp and normally given in kW for the electric motor. This chapter focuses on wheeled vehicles. The drag forces on wheeled vehicles are rolling resistance and aerodynamic drag. The engine and drive system for the vehicle must provide enough power to overcome the rolling resistance and aerodynamic drag on the vehicle.

Rolling Resistance [1]: The rolling resistance force comes from the interaction of the wheels with the pavement or driving surface. It is proportional to the weight of the vehicle. Other things equal, the rolling resistance force will double if we double the weight of the vehicle. Rolling resistance force increases a little as the speed of the vehicle increases, but the primary effect is the weight of the vehicle. The equation to calculate the rolling resistance force is:

$$F = (W)(Crr) \tag{2.1}$$

Where F is the rolling resistance force, W is the weight of the vehicle, and Crr is the rolling resistance coefficient. The rolling resistance coefficient is measured experimentally. For pneumatic tires rolling on pavement, Crr is a function of the type and size of tires used and the air pressure in the tires. This section will focus on using pneumatic tires on the pavement because that is the case for nearly all wheeled vehicles. When driving in sand or gravel or off-roading, the rolling resistance is much higher than on pavement. Trains have steel wheels rolling on steel rails, which is a lower rolling resistance than pneumatic tires on pavement. Trains are an efficient way to move heavy loads. We will talk about trains in a later section.

Pneumatic rubber tires have a lower rolling resistance than solid rubber tires or any other type of wheel we would consider using on cars, trucks, bicycles, motorcycles, scooters, etc. Companies that make tires are always looking for ways to improve the durability, performance, and rolling resistance. Rolling resistance is especially important for the tires used on large trucks. Lower rolling resistance tires allow the trucks to get better fuel economy, which lowers shipping costs and allows trucking companies to make more profit.

As a tire rolls on the pavement, the portion of the tire in contact with the pavement is flattened slightly in the area called the contact patch. The rolling motion causes the tire to be flexed back and forth as it goes through the process of being flattened and un-flattened. Rubber has a hysteresis loop and flexing the tire back and forth causes the tire to heat up. Some of the energy the vehicle uses is converted to heat in the tires because of the mechanics of rolling. The energy lost to heat in the tire is the most common explanation given for rolling resistance. The flattening and un-flatting of the tire also causes the rubber to slip a little on the pavement as it rolls. The rolling resistance is low with properly inflated tires but not negligible. It is a significant force and energy loss for wheeled vehicles, especially for heavy vehicles like trucks and trains.

Most of us rode bicycles when we were young and remember that, when the bicycle had a low tire, it took more effort to pedal the bicycle than when the tires were inflated. The same thing happens with car and truck tires, except it is more subtle because the engine will have plenty of power to overcome the increase in rolling resistance. Driving a car with a low tire will cause a lot of heat to be put into the tire as you drive and will cause the tire to overheat, fail, and blow out. It is important to keep the tires properly inflated. There are a few statements below that will help you understand why some tires have lower rolling resistance than other tires:

1. High-pressure tires have lower rolling resistance than lower pressure tires.
   The rolling resistance of a tire will decrease as the tire pressure is increased, even when the pressure is increased beyond the manufacturer's recommended

pressure. High-pressure tires are more susceptible to damage and do not have as good of performance in cornering, braking, and acceleration as lower pressure tires. Choosing the best tire for a vehicle involves more than just rolling resistance. However, generally, using higher pressure tires will lower the rolling resistance. Racing bicycles use high-pressure tires because they have a very low rolling resistance.

2. Narrow tires tend to have a lower rolling resistance than wider tires. When we turn a vehicle to go around a corner, the turn radius is slightly different for the inside and outside edges of the tires. The difference in turn radius causes scrubbing between the inside and the outside edge of the tire, increasing rolling resistance.

3. Large-diameter tires have lower rolling resistance than smaller diameter tires. This has to do with the geometry of flattening the contact patch. The deformation is more severe for small diameter tires than for large diameter tires, which results in less energy lost in the deformation and lower rolling resistance.

To quantify the energy and power required to overcome rolling resistance, we need typical values for the rolling resistance coefficient Crr. The table below lists typical values for different types of wheeled vehicles [1, 2, 3] (**Table 2.1**):

TABLE 2.1 Typical values for rolling resistance coefficient Crr.

Vehicle	Rolling Coefficient Crr
Racing Bicycle	0.004
Electrothon or Solar Car	0.0055
Motorcycle Street Bike. The narrower, higher-pressure tires have lower rolling resistance and the wider, high-performance tires have higher rolling resistance.	0.006–0.010
Large Truck (18-wheeler). The newer style tires have lower rolling resistance.	0.006–0.010
Passenger Cars and Trucks. The wider, high-performance tires tend to have higher rolling resistance.	0.010–0.012

© SAE International

These values can be used to estimate the rolling resistance at low speeds, but at higher speeds the rolling resistance coefficient increases, and we need a speed correction factor. Some textbooks define two rolling resistance coefficients, one constant and one multiplied by speed as illustrated below:

$$Crr = C_1 + (C_2)(V) \tag{2.2}$$

This approach is great from a theoretical point of view but, in practice, it is difficult to get an accurate experimental value for the C_1 term and almost impossible to get an accurate value for the C_2 term. Part of the increase in the rolling resistance coefficient comes from windage, which is an aerodynamic loss in the wheels, but is normally included with the rolling resistance. For this book, I will use an approximation that the rolling resistance coefficient varies with the term (1 + V/200) where the speed V has units of mph. With this assumption the rolling resistance force F_r can be calculated as:

$$F_r = (W)(Crr)\left(1 + \frac{V}{200}\right) \qquad (2.3)$$

Where W is the weight of the vehicle, Crr is the rolling resistance coefficient of the tires, and V is the speed of the vehicle in mph.

Example 2.1: Suppose a large truck weighs 70,000 lb and is traveling at 70 mph. What is the rolling drag force on the truck, and how much horsepower is required to overcome rolling drag? Use Crr = 0.006.

$$F_r = (70,000\,\text{lb})(0.006)\left(1 + \frac{70}{200}\right) = 567\,\text{lb} \qquad (2.4)$$

Power is the drag force multiplied by velocity. The power required to overcome rolling resistance for the truck is calculated as:

$$\text{Power} = (567\,\text{lb})(70\,\text{mph})\left(\frac{5280\,\text{ft}}{\text{mi}}\right)\left(\frac{\text{h}}{3600\,\text{s}}\right) = 58,212\,\text{ft-lb}/\text{s} = 105.8\,\text{hp} \qquad (2.5)$$

A loaded 18-wheeler truck has a typical weight of 70,000 lb. The engine will need to provide 105.8 hp to overcome rolling resistance on level ground at 70 mph. The engine will also have to overcome aerodynamic drag, which will be covered later in this chapter. When climbing a hill, the engine must provide the power to overcome gravitational loading. The horsepower required to climb hills can be higher than rolling and aerodynamic loading combined. We will cover gravitational loading in Unit 3.

Simplified Equations: In the United States, we will probably measure the weight of the vehicle in lb and the speed in mph. The data for the weight of vehicles and load limits on roads and bridges are usually given in pounds, kips, or tons. Speed limits are posted in mph, and most of us have a feeling for what the speed in mph means. To avoid having to go through the unit conversions every time, we will assume that the weight (W) of the vehicle is in lb and the speed (V) is in mph. Then, we can develop equations with the unit conversions all brought into one constant and simplify the analysis.

The drag force on a vehicle is normally measured in either lb or N. Eq. 2.6 and Eq. 2.7 below are used to get the drag force in units of lb or N. The power a vehicle needs to provide is normally given in W or kW for electric vehicles and horsepower for internal combustion vehicles. Eq. 2.8 and 2.9 below are used to get the power required in hp or W.

$$\text{Rolling Drag Force} = (W)(Crr)\left(1 + \frac{V}{200}\right)\text{lb} \qquad (2.6)$$

$$\text{Rolling Drag Force} = (4.459)(W)(Crr)\left(1 + \frac{V}{200}\right)\text{N} \qquad (2.7)$$

$$\text{Rolling Power} = (0.002667)(W)(V)(Crr)\left(1+\frac{V}{200}\right)hp \qquad (2.8)$$

$$\text{Rolling Power} = (1.989)(W)(V)(Crr)\left(1+\frac{V}{200}\right)W \qquad (2.9)$$

For emphasis, Eq. 2.6 to Eq. 2.9 above assume that the weight W is in pounds and the speed V is in mph.

Example 2.2: There is a small train ride at the mall for children. The engine weighs 280 lb and holds a 165 lb driver. Three cars weigh 140 lb each and carry up to 4 children. Assume that the average child weighs 40 lb. The track is 120-ft long on an oval loop. The train travels at 2.2 mph while running. The average rolling resistance is Crr = 0.02 (**Figure 2.1**).

    a. Calculate the energy required for the train to make one 120 ft loop around the track.

$$\text{Total Weight} = 280 + 165 + (140 + 4(40))(3) = 1345\,lb \qquad (2.10)$$

$$\text{Drag Force} = (1345\,lb)(.02)\left(1+\frac{2.2}{200}\right) = 27.2\,lb \qquad (2.11)$$

$$\text{Energy} = (27.2\,lb)(120\,ft) = 3263.5\,ft\text{-}lb \quad (\text{Energy per lap}) \qquad (2.12)$$

FIGURE 2.1   Small train.

b. Convert the energy to W-h.

$$(3263.5 \text{ ft-lb})\left(\frac{1.36 \text{ J}}{\text{ft-lb}}\right)\left(\frac{\text{W-h}}{3600 \text{ J}}\right) = 1.233 \text{ W-h} \quad (\text{Energy per lap}) \qquad (2.13)$$

c. Calculate the power the train uses while running in W.

$$\text{Power} = (1.989)(1345 \text{ lb})(2.2 \text{ mph})(.02)\left(1 + \frac{2.2}{200}\right) = 119 \text{ W} \qquad (2.14)$$

d. Based on this analysis, would one large deep-cycle lead-acid battery be adequate to power the train? Assume the battery has 1000 W-h of electric energy for powering the train.

$$\text{Time} = \frac{1000 \text{ W-h}}{119 \text{ W}} = 8.4 \text{ h} \quad (\text{continuous operation}) \qquad (2.15)$$

$$\text{Laps} = \frac{1000 \text{ W-h}}{1.233 \text{ W-h}/\text{lap}} = 811 \text{ laps} \qquad (2.16)$$

The answer is that a large deep-cycle lead-acid battery would be adequate to power the train.

Example 2.3: Suppose an adult person is comfortable walking at 3.5 mph for a long period. Assume that the power required for walking at this rate is 50 W, and the power varies linearly with the speed the person is walking.

Assume that the person is going to pull a wagon with children in it and the weight of the wagon and children is 120 lb. What is the walking speed pulling the wagon if the person is limited to 50 W? Assume the rolling resistance coefficient for the wagon is Crr = 0.015 (**Figure 2.2**).

From the information given in the problem, we can develop an equation relating the walking speed for the person to the power required. The power requirement is 50 W at 3.5 mph and varies linearly with the walking speed. This approach would be accurate over a range of speeds. If the person were to walk very slow or very fast, the power requirement would not vary linearly with speed. Over a range of speeds, the walking power for the person would vary according to the following equation:

$$\text{Walking Power} = (50 \text{ W})\left(\frac{V}{3.5 \text{ mph}}\right) \qquad (2.17)$$

The rolling power for the wagon is given by:

$$\text{Rolling Power} = (1.989)(120 \text{ lb})(V)(0.015)\left(1 + \frac{V}{200}\right) \text{W} \qquad (2.18)$$

FIGURE 2.2 Person pulling the wagon.

© SAE International

If we assume the person has 50 W of available power, the equation can be set up and solved as follows:

$$50\,\text{W} = (50\,\text{W})\left(\frac{V}{3.5\,\text{mph}}\right) + (1.989)(120\,\text{lb})(V)(0.015)\left(1 + \frac{V}{200}\right)W \quad (2.19)$$

Eq. 2.19 is solved to yield V = 2.79 mph. Using the assumptions given in the problem, the person would be able to pull the wagon at 2.79 mph with the same level of effort as walking at 3.5 mph. I am sure the reader recognizes that the person could swing their arms walking but would have to hold on to the wagon while pulling. The walking motion is a little different than the pulling motion and would impact the analysis. The simple analysis done in this problem is a good place to start and helps quantify why it takes more effort to pull the wagon than to walk, or that you must slow down when pulling the wagon if you want to maintain the same level of effort.

Aerodynamic Drag [4]: Aerodynamics and rolling resistance are the two large drag forces on ground vehicles rolling on level ground. Aerodynamics tends to be the more significant force for lightweight, high-efficiency vehicles like bicycles. Rolling resistance tends to be the dominant force for dense heavy vehicles like dump trucks. The aerodynamic drag force is given by the equation below:

$$\text{Aerodynamic Drag Force} = \tfrac{1}{2}\rho\,V^2\,A\,Cd \quad\quad\quad (2.20)$$

Where ρ = density of air, V = speed, A = Characteristic Area, and Cd is the drag coefficient. For vehicles, we normally use the frontal area for the characteristic area. The frontal area is the projected area seen while standing directly in front of the vehicle.

When doing the wind tunnel tests to measure Cd, it is possible to measure the density of the air, the velocity, and the drag force. A characteristic area is assumed, and the equation is used to calculate the drag coefficient. When looking up a drag coefficient in the literature, the number does not have any meaning unless you know the characteristic area used. Drag coefficients are sometimes calculated based on the total surface area of the vehicle, which yields a much different drag coefficient than when the frontal area is used. In this book, we will use the frontal area. Most drag coefficients for vehicles in the literature are based on the frontal area.

The power to overcome aerodynamic drag is the drag force multiplied by the speed:

$$\text{Aerodynamic Drag Power} = \tfrac{1}{2}\,\rho\, V^3\, A\, Cd \qquad (2.21)$$

Aerodynamic force is proportional to speed squared, but the power to overcome aerodynamic drag is proportional to speed cubed. At low speeds the aerodynamic power is small, but at high speeds the aerodynamic power becomes large. The aerodynamic drag coefficient Cd is related to the shape of the vehicle. The ideal shape to minimize aerodynamic drag is the teardrop or torpedo shape as illustrated in **Figure 2.3** below.

FIGURE 2.3 Airflow around teardrop shape.

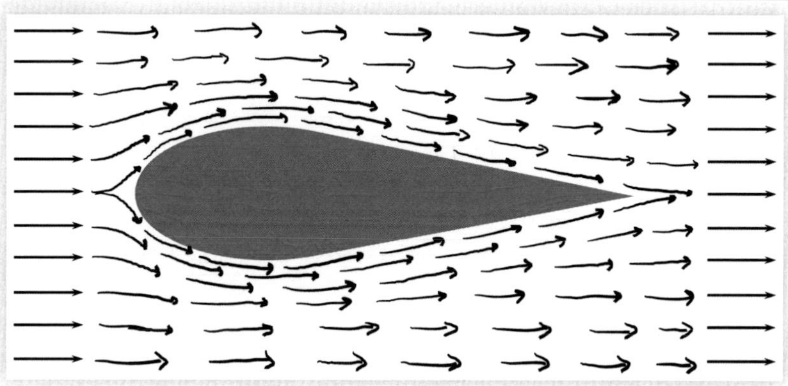

© SAE International

To minimize aerodynamic drag, the air should flow smoothly around the shape as illustrated in Figure 2.3. When the shape is less than ideal, the airflow streamlines separate from the object, which is often referred to as flow separation, as illustrated in **Figure 2.4** below.

FIGURE 2.4 Airflow around less ideal shapes.

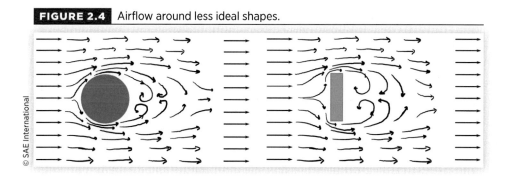

© SAE International

When air flows around a sphere or flat plate, the flow will separate from the object on the backside creating what are called eddy currents. Flow separation creates a low-pressure region behind the object that pulls backward on the object. There is a small increase in pressure on the front side of the object pushing backward, and a decrease in pressure behind the object pulling it backward. The lower pressure on the backside is the more significant term. Most of the drag on an object comes from the lower pressure, or suction pressure, on the backside. The teardrop shape prevents flow separation on the backside and greatly reduces the aerodynamic drag on the object.

These aerodynamic principles are used when designing the shape of most vehicles. The cars with the lowest drag coefficients tend to be rounded on the front and windshield and tapered on the back. There are many considerations in designing the shape of a car. Minimizing aerodynamic drag is important, but it would be impractical to taper the back of the car to a point. A compromise is made between aerodynamic drag, the functionality, drivability, and esthetics of the car.

For airplanes and jets, the aerodynamic drag is very important, and the shape of the fuselage and wings are designed to have a very low aerodynamic drag. The water drag on a ship hull is also a very important force, and ship hulls are designed to minimize water drag.

One way to look at the aerodynamic drag is to separate it into pressure drag and skin friction drag. The pressure drag comes from the airflow creating higher pressure on the front side of the vehicle and lower pressure on the backside of the vehicle. The drag force is then the difference between the front and back pressures multiplied by the frontal area of the vehicle. Pressure drag is the dominant term for cars and trucks, and that is why we use the frontal area as the characteristic area.

Skin friction drag is caused by the frictional (or shear) force between the object and the air. The drag force is proportional to the surface area of the vehicle. Skin friction is the dominant term for vehicles that have an excellent aerodynamic shape like airplanes and jets. Solar cars and electrathon vehicles are designed to have a low aerodynamic drag coefficient. When the vehicle has an excellent aerodynamic shape with very little flow separation, the dominant term will be skin friction drag, and it makes sense to use the surface area of the vehicle as to the characteristic area.

Example 2.4: Suppose the wind is blowing at 25 mph as you stand outside. The density of air is 1.225 kg/m³. You are 5' 6" (1.676 m) tall and your average width is 12 in (0.3048 m), making your frontal area 0.511 sq. m. Assume that your drag coefficient is Cd = 0.90. What is the force of the wind pushing on you? (or is it pulling on you? ☺) (**Figure 2.5**)

We are given all the information to make the calculation. The speed of 25 mph needs to be converted to 11.17 m/s, and then we plug into the formula for aerodynamic drag force:

$$\text{Aerodynamic Force} = \tfrac{1}{2}\rho\, V^2\, A\, Cd = \tfrac{1}{2}(1.225)(11.17)^2\,(0.511)(0.90) = 35.17\,\text{N}$$

(2.22)

FIGURE 2.5 Drag force caused by wind.

© Shutterstock

The 35.17 N answer can be divided by 4.448 to yield a force of 7.90 lb. Therefore, a 25-mph wind yields a drag force of about 8 lb, and we would notice an 8-lb force pushing on us, but it would not knock us over. At 40 mph wind speed, the force increases to 20.2 lb, which would be more noticeable. You would be leaning into the wind to keep from falling over. If the wind speed is increased to 70 mph, the drag force becomes 61.9 lb, which would make it hard to stand up. If it was a steady wind, you could probably lean into it and stand up.

We think of the wind pushing on us, but it is a little more accurate to think of the wind creating a low-pressure area behind us and suction pressure pulling us backward. There is an increase in pressure in front of us pushing us backward, but the low pressure behind us is the more significant term.

Airplane wings are tilted slightly in flight, so the air is directed downward slightly as it exits the rear of the wing. As air flows over an airplane wing there is a low-pressure region created above the wing and a high-pressure region created below the wing. The net lift on the wing is a combination of the two pressures, but the low pressure above the wing is the more significant term.

The examples and explanations given are to help the reader understand how aerodynamics work. When people first begin studying aerodynamics they recognized that the leading edge should be rounded to allow air to flow smoothly around the object. However, in the beginning, most people think that once the air gets past the object, it can no longer cause a drag force and they miss the part that the backside of the object should be tapered. Most of the drag on an object comes from the low-pressure region on the backside and tapering the backside to prevent flow separation is the most important part of reducing the drag on the object. Ideally, we round the leading edge and taper the back to minimize the drag on an object, but tapering the back is the most important part.

Now that we have a basic understanding of aerodynamics, we will need to develop equations to help us do energy and power analysis for cars and trucks. Recognizing that we will measure the speed of the vehicle in mph, we need to get more usable formulas for us to use in calculating the aerodynamic drag force and power.

The density of air varies a little with the temperature and pressure but is a fairly constant value. For this class, we will assume the density of air is 1.225 kg/m³.

In the literature, drag areas are typically given in square meters. As we use the equations we will assume that the frontal area of the vehicle is measured in square meters. The drag coefficient C_d is a dimensionless quantity.

The constants in the formulas include the ½ factor, the density of air, and all unit conversions necessary to allow us to measure the speed of the vehicle in mph and the frontal area in square meters. The equations below are similar to the equations used to calculate the rolling drag force and power, Eq. 2.6 to 2.9.

$$\text{Aerodynamic Drag Force} = (0.1199)\,V^2\,A\,Cd \quad N \tag{2.23}$$

$$\text{Aerodynamic Drag Force} = (0.02687)\,V^2\,A\,Cd \quad lb \tag{2.24}$$

$$\text{Aerodynamic Power} = (0.05357)\,V^3\,A\,Cd \quad W \tag{2.25}$$

$$\text{Aerodynamic Power} = (7.148 \times 10^{-5})\,V^3\,A\,Cd \quad hp \tag{2.26}$$

Where V is in mph, A is the frontal area in square meters, and Cd is dimensionless. For typical modern sedans, Cd is between 0.30 – 0.35. For typical SUVs, Cd is between 0.35 – 0.45 with more rounded SUVs having a lower drag coefficient. For pickup trucks, Cd is between 0.40 – 0.50. These values will be helpful allowing us to estimate the aerodynamic forces and power requirements of typical vehicles.

Example 2.5: Assume a typical sedan below is loaded with people and luggage so that the total weight is 4500 lb. The aerodynamic drag coefficient is 0.32, and the car will travel on a road trip where the car will average 70 mph. Assume a coefficient of rolling resistance of 0.010 for the tires.

a. Find the rolling drag force and the aerodynamic drag force (in pounds).

b. Find the rolling power and aerodynamic power (in horsepower).

c. If the gasoline engine and drive system in the car has an overall thermal efficiency of 25%, find the fuel economy of the car in mpg.

d. Suppose a diesel engine and drive system is used with an overall thermal efficiency of 30% and find the fuel economy of the car in mpg (**Figure 2.6**).

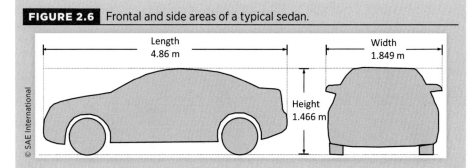

FIGURE 2.6 Frontal and side areas of a typical sedan.

© SAE International

The first order of business is to estimate the frontal area for the vehicle. There are many places on the internet to find the width and height of a vehicle, but the true frontal area is seldom given. A reasonable estimate is to assume the frontal area is 70% of the product for the width and height of the vehicle. The length of the car is not used in estimating the frontal area. To use the equations for aerodynamic drag force and power, the frontal area needs to be in square meters, and this vehicle is 1.466 m tall and 1.849 m wide. The frontal area is estimated as:

$$\text{Frontal Area} = (1.849\,\text{m})(1.466\,\text{m})(0.70) = 1.9\,\text{m}^2 \qquad (2.27)$$

The rolling drag force is calculated using Eq. 2.6, and the aerodynamic drag force is calculated using Eq. 2.24:

$$\text{Rolling Drag Force} = (4500)(0.010)\left(1+\frac{70}{200}\right) = 60.75\,\text{lb} \qquad (2.28)$$

$$\text{Aerodynamic Drag Force} = (0.002687)\left(70^2\right)(1.9)(0.32) = 80.05\,\text{lb} \qquad (2.29)$$

Rolling power is calculated using Eq. 2.28, and aerodynamic power is calculated using Eq. 2.29:

$$\text{Rolling Power} = (0.002667)(4500)(70)(0.010)\left(1+\frac{70}{200}\right) = 11.34\,\text{hp} \qquad (2.30)$$

$$\text{Aerodynamic Power} = \left(7.184\times10^{-5}\right)\left(70^3\right)(1.9)(0.32) = 15\,\text{hp} \qquad (2.31)$$

The total power required to overcome rolling power and aerodynamic power is 26.34 hp. The power will need to be provided by the engine. It is convenient to convert hp into units of BTU/s as we work through the calculations. We do not normally think of engine power in units of BTU/s but, while estimating fuel consumption and fuel economy, it is a convenient unit to use.

$$(26.32\,\text{hp})\left(\frac{550\,\text{ft-lb}/\text{s}}{\text{hp}}\right)\left(\frac{\text{BTU}}{778.2\,\text{ft-lb}}\right) = 18.62\,\text{BTU}/\text{s} \qquad (2.32)$$

The engine/drive system must provide 18.62 BTU/s to the wheels to overcome the rolling and aerodynamic drag on the car. The engine/drive system for the car is 25%

efficient, so the rate of fuel consumption required to get the 18.62 BTU/s power to the wheels is obtained by dividing by 0.25.

$$\text{Fuel Consumption} = \frac{18.62\,\text{BTU}\,/\,\text{s}}{0.25} = 74.46\,\text{BTU}\,/\,\text{s} \tag{2.33}$$

The gasohol fuel (10% ethanol) has 121,000 BTU/gallon energy, and there are 3600 s in an hour. The fuel consumption in gallons per hour (gph) is calculated as:

$$\text{Fuel Consumption} = \left(74.46\,\text{BTU}\,/\,\text{s}\right)\left(\frac{\text{gallon}}{121{,}000\,\text{BTU}}\right)\left(\frac{3600\,\text{s}}{\text{h}}\right) = 2.22\,\text{gph} \tag{2.34}$$

The fuel economy of the vehicle is then the speed (70 mph) divided by the fuel consumption (2.22 gph).

$$\text{Fuel Economy} = \frac{70\,\text{mph}}{2.22\,\text{gph}} = 31.6\,\text{mpg} \tag{2.35}$$

A gasoline-powered sedan would achieve a highway fuel economy of 31.6 mpg, which is typical for a midsized sedan. Part d of the problem is to investigate how powering the car with a diesel engine would impact the fuel economy. Diesel fuel has a higher energy content per gallon (138,700 BTU/gal), and the diesel engine will have higher thermal efficiency (30%). The combination of these two things causes the diesel-powered car to have a significantly higher fuel economy in miles per gallon. It still requires 26.32 hp = 18.62 BTU/s to power to car at 70 mph on level ground. The calculations for fuel economy of a midsized sedan are below:

$$\text{Fuel Consumption} = \frac{18.62\,\text{BTU}\,/\,\text{s}}{0.30} = 62.07\,\text{BTU}\,/\,\text{s} \tag{2.36}$$

$$\text{Fuel Consumption} = \left(62.07\,\text{BTU}\,/\,\text{s}\right)\left(\frac{\text{gallon}}{138{,}700\,\text{BTU}}\right)\left(\frac{3600\,\text{s}}{\text{h}}\right) = 1.61\,\text{gph} \tag{2.37}$$

$$\text{Fuel Economy} = \frac{70\,\text{mph}}{1.61\,\text{gph}} = 43.5\,\text{mpg} \tag{2.38}$$

This example illustrates the difference in using a diesel engine instead of a gasoline engine to power the car. The diesel-powered vehicle will get significantly better fuel economy than the gasoline-powered vehicle. Part of the reason for the better fuel economy is that diesel fuel has more energy per gallon than gasoline. The diesel engine will also have a higher thermal efficiency than the gasoline engine. Lost in this analysis is that the diesel engine will be heavier and more expensive than the gasoline engine. With the added weight, the diesel car will not get quite as good of fuel economy as in this example.

Example 2.6: Suppose a well-conditioned human can produce 70 W of useful power (about 0.1 hp). Suppose the person and bicycle together weigh 200lb, and the rolling resistance coefficient of the tires is 0.005. Assume the frontal area is 1 m², and the drag coefficient is 0.7.

What is the steady-state speed for the person on level ground? (**Figure 2.7**)

The person is capable of providing 70 W power, and the power will be used to overcome rolling and aerodynamic drag.

$$\text{Power} = \text{Rolling Power} + \text{Aerodynamic Power} = 70\ \text{W} \qquad (2.39)$$

$$(1.989)(200)(V)\left(1 + \frac{V}{200}\right) + (0.05357)\left(V^3\right)(1)(0.7) = 70 \qquad (2.40)$$

$$V = 10.8\ \text{mph} \qquad (2.41)$$

This analysis shows that the person would be able to travel along at 10.8 mph average speed. Notice that most of the power (47.3 of the 70 W) goes in overcoming the aerodynamic drag. Even at a moderate speed of 10.8 mph, most of the power goes in overcoming the aerodynamic drag. The aerodynamic drag power goes with the cube of speed so trying to go faster requires a lot more power. Going slower will require significantly less power.

Producing 70 W is a moderate output for an average person. If the equation is solved for 35 W power the average speed would be 7.92 mph. Producing 35 W is a relatively low power output for an average person. If the person is not in a big hurry, they can ride along at 8 mph without expending a lot of effort.

FIGURE 2.7   Person riding a bicycle.

© Shutterstock

A well-trained, young-adult man can produce about 200 W sustained power. Solving the equation for 200 W yields an average speed of 16.37 mph. At that speed 164.5 of the 200 W (82%) goes to aerodynamic power. In a bicycle race, the riders will lean over the handlebars to reduce their frontal area, reducing the aerodynamic power required. Reducing aerodynamic drag is the main consideration if you want to go fast on a bicycle. The fastest human-powered vehicles put the rider in a recumbent position to reduce the frontal area and use a fairing around the vehicle to reduce the aerodynamic drag coefficient.

Example 2.7: A team is preparing a solar car track race where the goal is to go as far as possible during an 8-h period. We will assume it is a large oval track, which, for the speeds of the solar cars, can be considered flat and level.

The rules of the race allow the car to carry batteries that will hold 3000 W-h of electric energy to power the car and a solar array that provides an average of 600 W during the 8-h period. The car and driver have a total weight of 600 lb. The tires used have a rolling resistance coefficient of 0.0055. The car has a frontal area of 1.2 m² and a drag coefficient of 0.105. The electric motor and power system have an overall efficiency of 94% in converting the energy from the batteries and solar array to the mechanical energy required to power the car.

Find the power (in W) required to drive the car at 25 mph. Using this value, how much energy will the car use during the 8-h day? Will there be enough energy in the batteries and from the solar array to drive all day at 25 mph? (**Figure 2.8**)

FIGURE 2.8 Solar car.

The total energy available is 3000 W-h in the batteries and 600 W for 8 h from the solar array, a total of 7800 W-h of energy.

$$\text{Energy Available} = 3000 \text{ W-h} + (600 \text{ W})(8 \text{ h}) = 7800 \text{ W-h} \qquad (2.42)$$

For the first part of the example problem, we will assume the car is to drive at 25 mph. We calculate the power for driving at 25 mph from the basic parameters of the vehicle.

$$\text{Power} = (1.989)(600)(25)(0.0055)\left(1 + \frac{25}{200}\right) + (0.05357)(25^3)(1.2)0.105 = 290 \text{ W}$$

$$(2.43)$$

The motor/drive system has an efficiency of 94%, so the power to be drawn by the motor is 290 W above divided by 0.94.

$$\text{Motor Power} = \frac{290 \text{ W}}{0.94} = 308.6 \text{ W} \qquad (2.44)$$

If we drive for 8 h, the total energy used for the day can be calculated as:

$$\text{Energy Used} = (308.6 \text{ W})(8 \text{ h}) = 2469 \text{ W-h} \qquad (2.45)$$

Since we have 7800 W-h of energy available for the day and only need 2469 W-h, the car will be able to drive all day. We will not run out of energy, but we will have a lot of energy left over at the end of the day. In solar car racing, the goal is to travel as far as possible during the day, so this would be a poor strategy for the race. We would only travel 200 mi. We should have driven faster and used up more of the energy. The subsequent examples help the reader understand how to work towards an optimum solution to racing the solar car.

Example 2.8: Find the power (in W) required to drive the solar car at 45 mph. Using this value, how much energy will the car use during the 8-h day? Will there be enough energy in the batteries and from the solar array to drive all day at 45 mph? If not, how far will the car have gone when the batteries are dead?

$$\text{Power} = (1.989)(600)(45)(0.0055)\left(1 + \frac{45}{200}\right) + (0.05357)(45^3)(1.2)0.105 = 977 \text{ W}$$

$$(2.46)$$

$$\text{Motor Power} = \frac{977 \text{ W}}{0.94} = 1039.3 \text{ W} \qquad (2.47)$$

$$\text{Energy Used} = (1039.3 \text{ W})(8 \text{ h}) = 8314 \text{ W-h} \qquad (2.48)$$

The 8314 W-h of energy required exceeds the 7800 W-h of energy available, so it is not possible to drive all day at 45 mph. At some point during the day the batteries will run out of energy, and the car will have to stop. Since the car stops before the 8-h period is complete, the solar array will not have time to gather the 4800 W-h of energy originally calculated. The equation below can be solved to find the time T when the car runs out of energy:

$$(1039.3\,\text{W})(T) = 3000\,\text{W-h} + (600\,\text{W})(T) \qquad T = 6.829\,\text{h} \qquad (2.49)$$

The car will need to stop with about 70 min remaining in the 8-h race period. It will travel a total of 308.3 mi when it runs out of energy ([45 mph] [6.829 h] = 308.3 mi). This is a much better strategy than driving at 25 mph all day and only traveling 200 mi, but it is not the optimum strategy. If this strategy were used, the car could sit parked beside the road for around 40 min charging the batteries and then drive for a while until the end of the 8-h driving period. The car would be able to go more than 307.3 mi, but this is still not the optimum strategy for racing.

Example 2.9: What is the ideal target speed for the solar car so it runs out of energy at the finish line and goes as far as possible on the energy available?

To get the optimum performance from the car we would want to drive steady all day and run out of energy at the end of the 8-h period.

$$(\text{Driving Power})(8\,\text{h}) = 7800\,\text{W-h} \qquad (2.50)$$

$$\frac{\left[(1.989)(600)(V)(0.0055)\left(1+\dfrac{V}{200}\right)+(0.05357)(V^3)(1.2)(0.105)\right](8)}{0.94} = 7800$$
$$(2.51)$$

$$V = 43.8\,\text{mph} \qquad (2.52)$$

The optimum strategy for the race is to drive at 43.8 mph. The car will use all of the energy available during the 8-h period and will travel 350.2 mi ([43.8 mph] [8 h] = 350.2 mi). For an energy-limited race like this, it is the best strategy to drive the car at a steady speed during the racing period.

The next series of examples are designed to illustrate some of the fundamental problems in developing electric cars and trucks. Electric vehicles have been around for more than 100 years, and there are many cases where electric vehicles are the best choice. Electric wheelchairs are good examples. It would not make sense to use an internal combustion engine for an electric wheelchair. Electric forklifts and other small electric vehicles are used in manufacturing and warehouses. Electric vehicles are clean, quiet, and have no emissions. The problem has always been the limited range of the vehicle before it needs to be recharged. The range is too small for cars and trucks. That limitation has been known for more than 100 years, and we are yet to find a solution. The lithium-ion batteries developed in the last few decades offer the best chance for electric cars and trucks, but even those batteries have their limitations.

Example 2.10: Suppose that our project goal is to design an electric car that can drive across the country on one charge. That is, we will charge the batteries and drive 2500 mi from Jacksonville, Florida to Los Angeles, California, without stopping to charge the batteries.

We decide to use lead-acid batteries because they are cheap and available compared to other types of batteries. We find a 12-V, lead-acid battery that has 110 amp-h of charge capacity and weighs 68.2 lb. This is the battery that we will use.

For a first approximation, assume that we will modify an F-350 Ford truck. Assume that the aerodynamic drag area is 4 m² and the drag coefficient is 0.5. We have found some special tires that have a rolling resistance coefficient of 0.008. We know it will have to be loaded with lots of batteries, so we will estimate the weight to be 10,000 lb as a first guess.

If the electric motor is 90% efficient, how much energy and how many batteries will be required to make the trip? Assume we will drive at 60 mph to reduce the aerodynamic losses (**Figure 2.9**).

$$\text{Power} = (1.989)(10,000)(60)(0.008)\left(1 + \frac{60}{200}\right) + (0.005357)(60^3)(4)(0.5) = 35,554 \text{ W} \quad (2.53)$$

$$\text{Motor Power} = \frac{35,554 \text{ W}}{0.9} = 39,504 \text{ W} \quad (2.54)$$

To drive 2500 mi at 60 mph takes 41.67 h of driving. The total energy required for the trip is the power multiplied by the time.

$$\text{Energy for Trip} = (39,504 \text{ W})(41.67 \text{ h}) = 1.646 \times 10^6 \text{ W-h} \quad (2.55)$$

FIGURE 2.9   Ford F-350 truck.

© Shutterstock

We will use 12-V lead-acid batteries that have a 110 amp-h capacity. We can estimate the energy stored in the battery by multiplying the voltage by the amp-h capacity. This is a high estimate for the battery. It will probably not be able to provide quite this much energy because the voltage will drop below 12 Volts as the energy is drawn out. If we draw the energy out slowly, the battery will be able to provide close to this much energy.

$$\text{Each Battery} = (12\,\text{volt})(110\,\text{amp-h}) = 1320\,\text{W-h} \qquad (2.56)$$

$$\text{Batteries Required} = \frac{1.646 \times 10^6}{1320} = 1247\,\text{batteries} \qquad (2.57)$$

At 68.2 lb per battery, this is 78,310 lb of batteries to provide the necessary energy to drive the 2500 mi. The F 350 truck will carry a lot of weight, but it will not carry 78,310 lb of batteries. Even if we could put this much weight on the truck it would greatly increase the rolling resistance, which would require even more batteries to make the trip. The lead-acid batteries are too heavy to make this work. It is not possible to design a vehicle that can travel 2500 mi across the country using lead-acid batteries.

Example 2.11: Suppose the truck above weighs 5000 lb unloaded, and we plan to load 5000 lb of batteries on it. Assume that we can only use 90% of the rated capacity of the batteries without damaging them. How far can we go at 60 mph before the batteries will need to be recharged?

Since the truck has a total weight of 10,000 lb, the power required to drive at 60 mph is 39,504 W, as calculated in the previous example. If we can only use 90% of the energy in a battery, the energy per battery can be calculated as follows:

$$\text{The energy in one battery} = (12\,\text{volt})(110\,\text{amp-h})(0.9) = 1188\,\text{W-h} \qquad (2.58)$$

The number of batteries in 5000 lb is:

$$\text{Number of Batteries} = \frac{5000\,\text{lb}}{68.2\,\text{lb / battery}} = 73\,\text{batteries} \qquad (2.59)$$

The amount of time the truck can be driven is the total amount of energy in the battery system divided by the 39,504 W required to power the truck.

$$\text{Drive Time} = \frac{(1188\,\text{W-h / battery})(73\,\text{battery})}{39,504\,\text{W}} = 2.195\,\text{h} \qquad (2.60)$$

$$\text{Distance} = (60\,\text{mph})(2.195\,\text{h}) = 132\,\text{mi} \qquad (2.61)$$

For this example, 50% of the weight of the vehicle was lead-acid batteries. Even so, with an extremely large amount of batteries, the vehicle can only travel 132 mi before needing a recharge. This example helps illustrate why electric cars powered by lead-acid batteries was never a viable solution.

Example 2.12: Suppose that instead of carrying 5000 lb of lead-acid batteries, we carry 5000 lb of lithium-ion batteries. Lithium-ion batteries have the highest energy density of any batteries commonly available. It would be a very expensive battery pack. We find that there is a battery available with a nominal voltage of 12.8V and 100 amp-h capacity. The battery weighs 28 lb and costs $1300.00. How far can the truck go on 5000 lb of lithium-ion batteries? How much will the 5000-lb battery pack cost? Assume that we can use 90% of the theoretical energy available in the batteries. The size of the batteries is 12.75" long, 6.5" wide, and 8.7" tall (**Figure 2.10**).

The power required is still 39,504 W since the truck still weighs 10,000 lb. The energy stored on one battery is:

$$\text{The energy in one battery} = (12\,\text{volt})(100\,\text{amp-h})(0.9) = 1152\,\text{W-h} \qquad (2.62)$$

$$\text{Number of Batteries} = \frac{5000\,\text{lb}}{28\,\text{lb / battery}} = 178\,\text{batteries} \qquad (2.63)$$

$$\text{Drive Time} = \frac{(1152\,\text{W-h / battery})(178\,\text{battery})}{39,504\,\text{W}} = 5.19\,\text{h} \qquad (2.64)$$

$$\text{Distance} = (60\,\text{mph})(5.19\,\text{h}) = 311\,\text{mi} \qquad (2.65)$$

$$\text{Battery Cost} = (178)(\$1300.00) = \$231,400 \qquad (2.66)$$

FIGURE 2.10 | Lithium-ion battery.

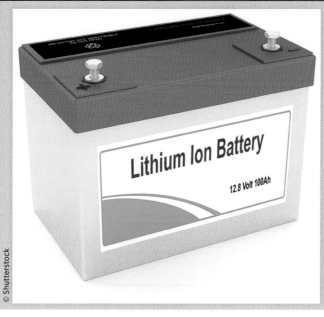

© Shutterstock

The range of 311 mi is much closer to a reasonable driving range that would be acceptable to most people than the 132 mi possible with the lead-acid batteries. If the batteries can be charged quickly, most people would be satisfied with a range of 311 mi. But, if it takes hours to charge the batteries, the truck is not suitable for traveling on the highway. Most people will want to travel more than 311 mi in a day. The cost of purchasing the batteries is much too expensive to be practical. The cost will come down as the manufacturing processes are improved. Lithium-ion batteries do not have precious metals or other materials that make them inherently expensive. It is reasonable to think that the cost of the batteries can be reduced to where it is reasonable.

Example 2.13: Suppose that instead of carrying 5000 lb of batteries, we carry 5000 lb of diesel fuel. If the diesel engine is 30% efficient at converting the chemical energy in the fuel to useful mechanical energy, how far can the truck go on 5000 lb of fuel?

$$\text{Power} = (1.989)(10,000)(60)(0.008)\left(1+\frac{60}{200}\right)+(0.05357)(60^3)(4)(0.5) = 35,554\,\text{W} \tag{2.67}$$

$$\text{Fuel Power} = \left(\frac{35,554\,\text{W}}{0.3}\right)\left(\frac{\text{BTU}}{1055\,\text{J}}\right)\left(\frac{3600\,\text{s}}{\text{h}}\right) = 404,401\,\text{BTU}/\text{h} \tag{2.68}$$

The diesel fuel has 138,700 BTU per gallon, and the density of the fuel is 7.3 lb/gallon. 5000 lb of fuel is (5000)/(7.3) = 685 gal. The energy stored in the 685 gal of fuel is 95×10^6 BTU. The time the truck can travel is then:

$$\text{Drive Time} = \frac{95 \times 10^6}{404,401} = 234.9\,\text{h} \tag{2.69}$$

$$\text{Distance} = (60\,\text{mph})(234.9\,\text{h}) = 14,095\,\text{mi} \tag{2.70}$$

With 5000 lb of diesel fuel (685 gallons), the truck could drive across the country five times on one tank of fuel. This series of examples should help the reader understand the largest challenge of changing to electric vehicles. It is impossible to make an electric vehicle that has a driving range comparable to vehicles powered by internal combustion engines. Electric vehicles are better than internal combustion powered vehicles in many ways. They are clean, quiet, and pleasant to drive. It's been that way for over 100 years, and we have yet to find a solution. Lithium-ion batteries are much better than lead-acid batteries, but they are still only marginally acceptable. We are still looking for a better solution.

2.3 Trains

Trains are wheeled vehicles but need to be analyzed differently from other wheeled vehicles. Trains are very heavy, and the rolling drag is much greater than the aerodynamic drag. It is reasonable to neglect the aerodynamic drag for typical trains. The engine for a light passenger train will typically weigh 120 tons (240,000 lb). A large freight-train engine will typically weigh 250 tons (500,000 lb). Passenger cars typically weigh 40 tons empty and 50 tons loaded. Freight cars vary considerably in weight and can be much heavier than passenger cars.

The pure rolling-resistance coefficient for steel wheels rolling on steel rails can be less than 0.001, but that is an idealistic situation. Rolling resistance increases with dirt on the track, sponginess of the ties underneath the rails, contact between the wheels and side of the rails, and corners. A realistic average value for the rolling resistance coefficient is 0.005 to 0.006 for trains, which is slightly better than the best large truck tires. Trains are the most energy-efficient way to move heavy loads over long distances.

Assume a 250-ton engine pulling 30 100-ton cars at 60 mph for 1200 mi. The average rolling resistance coefficient is 0.006 for the train. The diesel locomotive is 35% efficient in converting the energy in the diesel fuel into useful energy to power the train. Neglect aerodynamic drag. How much horsepower is required to pull the train at 60 mph? How much fuel is required to make the 1200 mi journey?

$$\text{Weight} = \left(250\,\text{ton} + 30(100\,\text{ton})\right)\left(2000\,\frac{\text{lb}}{\text{ton}}\right) = 6.5 \text{X}\, 10^6\,\text{lb} \tag{2.71}$$

For this analysis, we will use the horsepower required to pull the train along on the level ground. The engine will need more horsepower to pull the train upgrades, but the level ground approach is best for estimating the total energy required to make the 1200 mi trip. Eq. 2.8 is used to estimate the rolling power required. Aerodynamic drag is neglected for this analysis:

$$\text{Power} = (0.002667)(6.5 \times 10^6)(60(0.006)\left(1 + \frac{60}{200}\right) = 8113\,\text{hp} \tag{2.72}$$

$$\text{Time for 1200 mi} = \frac{1200\,\text{mi}}{60\,\text{mph}} = 20\,\text{h} \tag{2.73}$$

The total energy required from the engine is the average power multiplied by the travel time.

$$\text{Energy} = (8113\,\text{hp})(20\,\text{h}) = 162,260\,\text{hp-h} \tag{2.74}$$

The energy unit of hp-h is a commonly used unit for farm tractors and trains. The diesel engine used for the train is 35% efficient in converting the chemical energy in the fuel into useful energy. Fuel energy is calculated as follows:

$$\text{Fuel Energy} = \frac{162,260\,\text{hp-h}}{0.35} = 463,601\,\text{hp-h} \tag{2.75}$$

To get fuel consumption, we need to convert the hp-h into BTU and recognize that diesel fuel has 138,700 BTU per gallon. One hp-h of energy is equal to 2544.3 BTU.

$$\left(463,601\,\text{hp-h}\right)\left(\frac{2544.3\,\text{BTU}}{\text{hp-h}}\right)\left(\frac{\text{gal}}{138,700\,\text{BTU}}\right)=8504\,\text{gal} \tag{2.76}$$

The train will use 8504 gal of diesel fuel to make the 1200 mi trip. The fuel economy is 0.141 mpg, but it is more relevant to measure the ton-miles per gallon for a train. The ton-miles per gallon accounts for the fact that the train is moving a large amount of freight over the distance.

$$\frac{\left(3250\,\text{ton}\right)\left(1200\,\text{mi}\right)}{8504\,\text{gal}}=459\,\text{ton-mi / gal} \tag{2.77}$$

2.3.1 Summary for Wheeled Vehicles

The drag force on wheeled vehicles comes from rolling drag and aerodynamic drag. For lightweight vehicles such as bicycles, motorcycles, and scooters the aerodynamic drag force is the dominant term. These vehicles have a poor aerodynamic shape, which gives them a high aerodynamic drag force for their size. Rolling and aerodynamic drag are both significant factors for cars and trucks. Rolling power tends to be more dominant at lower speeds, and aerodynamic power tends to be more dominant at higher speeds.

To get the power to the wheels to overcome rolling and aerodynamic drag, the energy must go through the internal combustion engine or electric motor and the drive system. To calculate the fuel power or electric power required we divide the total of rolling and aerodynamic power by the efficiency of the motor/drive system.

$$\text{Fuel Power or Electric Power} = \frac{\text{Rolling} + \text{Aerodynamic Power}}{\text{motor \& drive efficiency}} \tag{2.78}$$

The motor/drive efficiency includes the efficiency in the drive system. Significant power is lost in the transmission and final drive systems of a conventional car or truck. The overall efficiency of a conventional gasoline-powered car or truck is about 25%. Diesel-powered cars and trucks are closer to 30% efficient, and the best large 18-wheeler trucks have an overall efficiency that can approach 35%. These efficiency values are for highway driving. In the city driving, the thermal efficiency of the internal combustion engine is much lower than on the highway.

The motor/drive efficiency for electric cars is typically 85%. Electrathon and solar cars can have efficiencies that approach 95%. Bicycle drive systems are essentially 100% efficient. The human powering the bicycle is typically 20% efficient in converting the carbohydrate energy into useful energy to power the bicycle.

This unit provided the tools for analyzing wheeled vehicles and estimating the energy consumption and overall efficiency of the vehicle. For further depth, the reader should work through the homework problems below.

2.4 **Homework**

1. Your shipping company manages. a small fleet of 18-wheeler trucks. The tires on the trucks have an average rolling resistance coefficient of 0.0075. There is the opportunity to use a new and improved tire design that will have a rolling resistance coefficient of 0.0072. Assume that the average truck in your fleet weighs 60,000 lb and that you average 5,000,000 mi per year in truck trips. Also, assume that on average the trucks are traveling at 65 mph. Assume that the efficiency of the diesel engine and drive system is 35%.

    a. How much energy per year would you save by going with the new and improved tires? Please express your answer in ft-lb and BTUs.

    b. Assuming an engine efficiency of 35%, how many gallons of fuel per year would be saved by going to the lower rolling resistance tires? (Answer 16,670 gallons)

2. Assume a typical pickup below is loaded so that the total weight is 7000lb. The aerodynamic drag coefficient is 0.52, and the truck is to travel on a road trip where it will average 70 mph. Assume a coefficient of rolling resistance of 0.010 for the tires.

    a. Find the rolling drag force and the aerodynamic drag force (in pounds)

    b. Find the rolling power and aerodynamic drag power (in horsepower)

    c. If the gasoline engine and drive system in the truck has an overall thermal efficiency of 25%, find the fuel economy of the truck in miles per gallon.

    d. Suppose a diesel engine and drive system is used with an overall thermal efficiency of 30% and find the fuel economy of the truck in miles per gallon. (Answer 21.98 mpg) **(Figure 2.11)**

FIGURE 2.11 Dimensions for a typical truck.

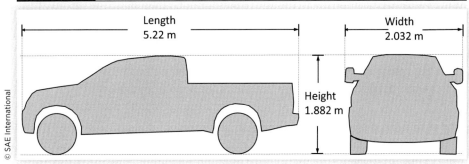

3. A team is preparing an electrathon vehicle for a contest where the goal is to complete a 100 km race as quickly as possible. The race will be conducted on an oval track which, for the speeds, the electrathon vehicle will travel, can be considered flat and level racing.

    The rules of the race allow the car to carry batteries that will hold 1000 W-h of electric energy to power the car. The car and driver will have a total weight of 500lb. BMX tires are used which have a rolling resistance coefficient of 0.006. The car has a frontal area of 0.8 m² and a drag

© SAE International

coefficient of 0.12. The electric motor and power system have an overall efficiency of 92% in converting the energy in the batteries to the mechanical energy required to power the car.

Find the power (in W) required to drive the car at 30 mph. Using this value, how much energy will be required for the car to complete the full 100 km distance of the race? Will there be enough energy in the batteries to allow the car to complete the race at 30 mph? (Answer Power = 374.7 W) (**Figure 2.12**)

FIGURE 2.12 Electrathon vehicle.

4. A team is preparing an electrathon vehicle for a contest where the goal is to complete a 100-km race as quickly as possible. The race will be conducted on an oval track, which for the speeds the electrathon vehicle will travel can be considered flat and level racing.

The rules of the race allow the car to carry batteries that will hold 1000 W-h of electric energy to power the car. The car and driver will have a total weight of 500lb. BMX tires are used which have a rolling resistance coefficient of 0.006. The car has a frontal area of 0.8 m² and a drag coefficient of 0.12. The electric motor and power system have an overall efficiency of 92% in converting the energy in the batteries to the mechanical energy required to power the car.

What is the optimized speed for the car to allow it to complete the full 100 km in the least time possible, using all energy in the batteries? Assume the car travels at constant speed the entire distance. What is the race time for the 100 km race? (Answer 38.65 mph)

5. A team is preparing an electrathon vehicle for a contest where the goal is to complete a 100 km race as quickly as possible. The race will be conducted on an

oval track, which for the speeds the electrathon vehicle will travel can be considered flat and level racing.

The rules of the race allow the car to carry batteries that will hold 1000 W-h of electric energy to power the car. The car and driver will have a total weight of 500lb. BMX tires are used which have a rolling resistance coefficient of 0.006. The car has a frontal area of 0.8 m² and a drag coefficient of 0.12. The electric motor and power system have an overall efficiency of 92% in converting the energy in the batteries to the mechanical energy required to power the car.

Suppose the team was too conservative in the beginning, traveling at 20 mph for the first 30 km of the race and realizing they should have been going faster. How much energy have they used in traveling the first 30 km of the race? How much energy is left in the batteries? How fast should they travel during the last 70 km of the race to get to the finish line as quickly as possible? (Answer 44.45 mph)

6. While testing an electrathon vehicle, it is observed that it uses 700 W when traveling at 42 mph. The vehicle weights 525 lb, and the tires have a rolling resistance coefficient of 0.0055. The electric motor has an efficiency of 95%. The frontal area of the vehicle is estimated to be 1.16 m². From this information, calculate the aerodynamic drag coefficient for the vehicle. (Answer 0.0810)

7. Assume that a person plans to ride an electric bicycle across the country, a distance of 3500 mi. The person can provide an average of 50 W of power to the pedals while riding. The battery on the bicycle has a useful capacity of 800 W-h energy. The person schedules 10 h per day for riding, but there will be rest breaks so that the actual riding time will be 8 h per day. Each night the person will charge the battery and then use the battery energy the next day. Assume that the electric motor is 85% efficient in converting the energy in the battery to useful mechanical energy. The rider and bicycle and battery have a total weight of 250 lb, and the big soft tires on the bicycle have a rolling resistance coefficient of 0.008. The frontal area of the bicycle and rider is 0.60 m² and the aerodynamic drag coefficient is 0.75.

   a. What is the average power available while riding the bike (human + battery)?

   b. Using the average power, what is the steady-state speed on level ground?

   c. Based on the steady-state speed, how many days to make the 3500 mi trip? (Answer 30.21 days)

How many Cal will the rider burn to power the bicycle on the 3500 mi trip? Please assume that the human muscles are 20% efficient in converting the carbohydrate energy into useful work. Assuming 3500 Cal per pound, how many pounds does this correspond to? (**Figure 2.13**)

FIGURE 2.13 Cross country bicycle trip.

© Shutterstock

8. An 80,000-lb truck is driving westward on I-70 across Kansas at a steady speed of 70 mph. Assume that the rolling resistance coefficient of the tires is 0.007, the frontal area of the truck is 7.8 m², and the aerodynamic drag coefficient of the truck is 0.62.

a. Find the horsepower required to push the truck along at steady speed on level ground.

b. Assuming the thermal efficiency of the diesel engine and transmission is 32%, find the gallons per hour fuel consumption of the truck.

c. Calculate the amount of fuel required for the truck to travel 400 mi. If the driver paid $2.15 per gallon, what are the fuel cost per mile for the truck and total fuel cost for the trip? (Answer $183.35) (**Figure 2.14**)

FIGURE 2.14 18-Wheeler truck.

© Shutterstock

9. We decide to use a small gasoline engine powered tractor to drag a log through the field to the house where it can be chopped and split for firewood.

   a. The log has a total weight of 800 lb. To guarantee we will have the power required to pull it, we make a high estimate of 0.9 for the coefficient of friction between the logs and the ground. Find the power required to drag the log at 10 kph. Neglect air resistance.

   b. Assume that for driving through the field the rolling resistance coefficient for the tractor is 0.1 and that the tractor has a weight of 3,000 lb. Find the power required to drive the tractor at 10 kph pulling the logs (i.e. add the rolling resistance power for the tractor to the power in an above). Neglect air resistance. (Answer 12.72 kW) (**Figure 2.15**)

FIGURE 2.15   Tractor pulling a log.

© Shutterstock

10. An electric vehicle is to be developed to allow people to drive around inside a large warehouse facility. The vehicle will average 5 mph speed while driving and the rolling resistance coefficient for the tires is 0.020. The electric motor is 85% efficient in converting the electrical energy stored in the batteries to useful mechanical energy. Air resistance is negligible for this problem. The ten deep-cycle lead-acid batteries that power the vehicle weight 65 lb each and have a useful energy capacity of 1kW-h each. The vehicle weighs 250 lb without batteries, people, or cargo. The vehicle is designed to carry up to 500 lb of people and cargo load not counting the batteries.

a. What is the total weight of the vehicle plus batteries plus a maximum load of people and cargo?

b. What percentage of the total weight (above in a) comes from the batteries?

c. Assuming the vehicle is fully loaded with 500 lb of people and cargo, calculate the power in W that must be drawn from the batteries to power the vehicle along at 5 mph. (Answer 335.8 W)

d. Based on the total energy capacity of the batteries, how long can the vehicle travel around the warehouse before it needs to have the batteries recharged? (Answer 29.8 h) (**Figure 2.16**)

FIGURE 2.16   Small electric vehicle.

© Shutterstock

11. Assume a 220-ton engine is pulling 18 cars that weigh 110 tons each at 55 mph for 2000 mi. The average rolling resistance coefficient is 0.005 for the train. The diesel locomotive is 34% efficient in converting the energy in the diesel fuel into useful energy to power the train. Neglect aerodynamic drag. How much horsepower is required to pull the train at 55 mph? How much fuel is required to make the 2000-mi journey? (Answer 8075 gallons) (**Figure 2.17**)

FIGURE 2.17   Train.

© Shutterstock

References

1.   Carroll, *The Winning Solar Car*, SAE, 2003.

2.   https://www.engineeringtoolbox.com/rolling-friction-resistance-d_1303.html

3.   https://en.wikipedia.org/wiki/Rolling_resistance

4.   Munson et al, *Fundamentals of Fluid Mechanics*, John Wiley & Sons, 1994. Or any good fluid mechanics textbook.

Boats and Ships

3.1 Drag Force for Boats and Ships

The drag force on a boat or ship is more complicated than for an object moving within a fluid because the boat is riding on the surface of the water. There exists a normal friction drag on the boat hull due to the water flowing against the hull. A poorly designed hull can produce a significant amount of pressure drag on the hull pushing the water up in front of the boat and allowing the flow to separate on the back of the boat. The drag force from water flow around the hull is the dominant drag term. The portion of the boat or ship above the water will experience aerodynamic drag, which is significant but smaller in magnitude than the water drag force.

Boats and ships create waves as they pass through the water, and it takes a significant amount of energy to create the waves. This was studied extensively by William Froude, and he developed the Froude number that goes with the drag on the boat or ship. The Froude number was found to be linearly proportional to the speed of the ship in the water and inversely proportional to the square root of the length of the ship. Froude's original work did not include the acceleration of gravity in the definition of the Froude number, but gravity was added later to make the Froude number a dimensionless quantity as is common in the subject of fluid mechanics. The modern definition for the Froude number (Fr) is below [1]:

$$Fr = \frac{V}{\sqrt{gL}} \tag{3.1}$$

Where V is the speed of the boat in the water, g is the acceleration of gravity, and L is the length of the boat or ship at the waterline. The Froude number is a dimensionless quantity, that is, the units divide out. Boats and ships with similar Froude numbers will create similar waves, assuming a good hull design. The hull should be designed to allow water to flow smoothly around the boat or ship.

3.1.1 A Brief Derivation of the Froude Analysis for Ships

It has been known for at least 150 years that the drag force on a ship is proportional to the weight of the ship and the square of the Froude number. The reader could accept this as fact and skip this derivation. The derivation is included in the book to give the reader a deeper understanding of where the Froude analysis comes from. The basic water drag force on a ship is given by the equation below:

$$\text{Drag Force} = \frac{1}{2}\rho V^2 A\, Cd \qquad (3.2)$$

Where ρ is the density of water, V is the speed of the ship in the water, A is the surface area of the hull in contact with the water, and Cd is the drag coefficient of the hull (based on the surface area of the hull in contact with the water).

For a ship of weight mg, the volume of water displaced must have a weight equal to the weight of the ship. The volume of water displaced is given by the following:

$$\text{Volume Displaced} = \frac{mg}{\rho g} = \frac{m}{\rho} \qquad (3.3)$$

Where m is the total mass of the ship and ρ is the density of water. Seawater has a higher density than freshwater.

The assumption made in the Froude analysis is that there is a relationship between the surface area of the hull in contact with the water (A) and the volume of water displaced by the hull (volume displaced). A shape factor (SF) is defined so that the area (A) is proportional to the volume divided by the length of the hull in contact with the water (L). The equation for the assumption is shown below:

$$A = (SF)\frac{(\text{Volume Displaced})}{L} = (SF)\frac{\left(\dfrac{m}{\rho}\right)}{L} = (SF)\left(\frac{m}{\rho L}\right) \qquad (3.4)$$

Equation 3.4 would be true for any given ship, but the equation would be of little value if the SF constant is different for different sizes of ships. Equation 3.4 and the Froude analysis are only valuable if the SF is constant for ships that have a similar hull shape.

It is difficult to calculate the SF for real hulls, but we can get an idea of the correct value by looking at an idealized shape. The shape to be considered is the hemisphere. Suppose the hull of the ship in question is a hemisphere. A more precise way of saying this is that the area of the ship in contact with the water is a hemisphere. The area of a hemisphere is $A = 2\pi r^2$, the volume is given by Volume $= (2/3)\pi r^3$, and the length is $L = 2r$. The shape factor for a hemisphere is 6. That is:

$$SF = \frac{AL}{\text{Volume}} = \frac{\left(2\pi r^2\right)\left(2r\right)}{\dfrac{2}{3}\pi r^3} = 6 \qquad (3.5)$$

The SF is a constant for all sizes of hemispheres. It does not vary with the radius of the hemisphere. While this is not a proof, it can be shown mathematically that the SF is constant for a given hull shape. Once a hull shape is defined, making the hull larger or smaller does not change the SF. Different hull shapes will have SF, but if we assume the ship has a good hull shape, the SF will be fairly constant from ship to ship.

Equation 3.4 is substituted into Eq. 3.2 to yield the following:

$$\text{Drag Force} = \frac{1}{2}\rho\,V^2\,(SF)\left(\frac{m}{\rho L}\right)Cd \qquad (3.6)$$

The constants can be gathered at the beginning of the equation. Using the definition of the Froude number in Eq. 3.1, Eq. 3.6 can be rewritten as:

$$\text{Drag Force} = \left[\left(\frac{1}{2}\right)(SF)(Cd)\right]mg\,Fr^2 \qquad (3.7)$$

Equation 3.7 says that the drag force on a ship will be linearly proportional to the weight of the ship (mg) and proportional to the square of the Froude number. The constant term is determined empirically.

A detailed discussion of the hydrodynamics of boats is beyond the scope of the class. The water drag force on a boat or ship is proportional to the weight of the boat or ship, so in some ways, it is similar to rolling resistance on a wheeled vehicle. For lower speeds the water drag force increases with the square of the speed, making it similar to aerodynamic drag. As the boat speeds up it begins to plane on top of the water and the drag force tends to increase linearly with speed. There is a transition region as the boat approaches and surpasses the hull speed, as defined below:

$$\text{Hull Speed} = V_{Hull} = (0.3985)\sqrt{gL} \qquad (3.8)$$

The model below will be used to estimate the water drag force on boats and ships in this book. The model assumes the boat or ship has a good hydrodynamic shape so that the water flows smoothly over the hull. For barges with square noses and sterns, the model will underestimate the drag force.

1. Calculate the Froude Number. The first step is to calculate the Froude number for the boat or ship. In the literature, it is common to list the length of the boat or ship but not common to list the waterline length. The waterline length is typically about 90% of the length, but you may need to use your judgment for the boat or ship in question.

2. Calculate Hull Speed (Eq. 3.8). The hull speed is the speed where the bow and stern of the boat or ship are riding on the crests of the transverse waves generated by the boat or ship. The waterline length of the boat or ship is equal to the wavelength of the waves generated when the boat or ship is traveling at hull speed. The hull speed is a limiting speed for most human-powered boats and large ocean ships. Achieving hull speed requires a large amount of power compared to traveling at 85% of hull speed. Olympic athletes, in specially designed boats, can exceed the hull speed but for most human-powered boats the human cannot provide enough power to achieve hull speed. For large ships, the power to achieve hull speed is so high that it

is impractical and cost-prohibitive for ships to travel at the hull speed. Motorboats and jet skis routinely travel much faster than the hull speed because they have the horsepower to do it.

3. $0 < V < (0.85)\, V_{hull}$ is low speed operation. Drag is proportional to the weight of the boat or ship and the Froude number squared. Most of the ships and boats used for transportation of freight and people operate in this range of speeds. It is the most important part of the model as far as energy consumption and energy efficiency of boats and ships.

$$\text{Drag Force} = (0.113)\text{mg}\left(\text{Fr}^2\right) \tag{3.9}$$

$$\text{Drag Power} = (\text{Drag Force})\,V \tag{3.10}$$

If the ship or boat is operating at a speed below 85% of the hull speed, Eq. 3.9 and Eq. 3.10 gives a good estimate of the drag force and power required. For speeds above 85% of hull speed, the boat will begin to rise in the water and plane out. As the boat planes out the shape factor in Eq. 3.5 changes significantly. The mass of water displaced by the hull is less than the mass of the boat. The assumptions made in deriving Eq. 3.9 and Eq. 3.10 are invalid, and the equations over-estimate the drag force and power required.

Motorboats operate at speeds well above the hull speed. The analysis below is a method for estimating the drag force and power requirement for boats operating at speeds above 85% of the hull speed. Boat hulls are designed to operate over a range of speeds. We will assume that the boat is being operated within the speed range for which it was designed. There is a relationship between the weight of the boat, the speed, and the shaft horsepower. This relationship is shown graphically in **Figure 3.1**.

Graphical estimations can be made using Figure 3.1. If any two of the parameters are known, the third can be estimated from the figure by drawing a line connecting the two known parameters. The example in the figure is for a boat that has a weight of 3000 lb and is traveling at a speed of 36 knots. From the graph, the engine will need to deliver 90 hp to the shaft driving the propeller to push the boat along. The analysis assumes the boat is using a propeller, or propellers, that have been designed for the boat and speed. Everything about the analysis assumes that the boat and power system have been properly designed for the operating speed and that the boat is cruising at a constant speed.

Some readers may want to program this analysis into the computer. The graphical method using Figure 3.1 does not lend itself to programming. The algebraic method below will be useful when programming the analysis.

1. Assume the weight of the boat (W) is known in lb, and the speed (S) is known in knots. The horsepower can be calculated using the equations below:

$$x_D = 3.9447\left((\ln W) - 6.64158\right) \tag{3.11}$$

$$x_S = 4.21266\left(\left(\ln\frac{1}{S}\right) + 6.45673\right) \tag{3.12}$$

$$x_P = 2x_S - x_D \tag{3.13}$$

$$\text{Shaft Power} = 11{,}312\left(0.777625\right)^{x_P}\ \text{hp} \tag{3.14}$$

FIGURE 3.1   Planing Hull Speed, Displacement, and Shaft Horsepower Relationship

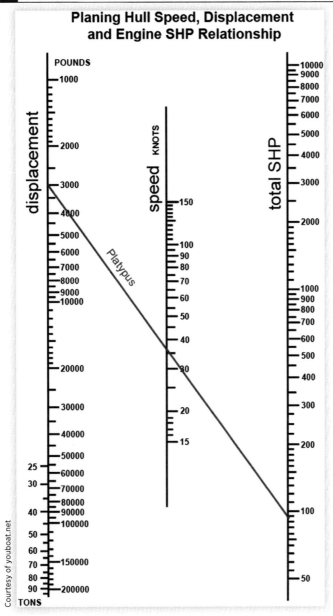

If the weight of the boat is 3000 lb, and the speed is 36 knots, calculations are: $x_D = 5.384$, $x_S = 12.10$, $x_P = 18.82$, Shaft Power = 99.4 hp.

2. Assume the weight of the boat (W) is known in lb, and the shaft horsepower (P) is known in hp. The speed of the boat can be calculated using the equations below:

$$x_D = 3.9447\big((\ln W) - 6.64158\big) \tag{3.15}$$

$$x_P = 3.97597\left(\left(\ln\frac{1}{P}\right) + 9.333586\right) \tag{3.16}$$

$$x_S = \frac{x_D + x_P}{2} \tag{3.17}$$

$$\text{Speed} = 636.977 (0.788692)^{x_S} \text{ knots} \tag{3.18}$$

If the weight of the boat is 3000 lb, and the shaft horsepower is 100 hp, calculations are: $x_D = 5.384$, $x_P = 18.80$, $x_S = 12.09$, Speed = 36.1 knots.

3. Assume the shaft horsepower (P) is known in hp, and the speed of the boat is known in knots. The weight of the boat can be calculated using the equations below:

$$x_P = 3.97597 \left(\left(ln \frac{1}{P} \right) + 9.333586 \right) \tag{3.19}$$

$$x_S = 4.21266 \left(\left(ln \frac{1}{S} \right) + 6.45673 \right) \tag{3.20}$$

$$x_D = 2x_S - x_P \tag{3.21}$$

$$\text{Weight} = 766.301 (1.28853)^{x_D} \text{ lb} \tag{3.22}$$

If the shaft horsepower is 100 hp, and the speed of the boat is 36 knots, calculations are: $x_P = 18.80$, $x_S = 12.10$, $x_D = 5.41$, Weight = 3018 lb.

Please recognize that the model above gives an estimate for the water drag force on the hull of a boat or ship. It does not include aerodynamic drag on the portion above the water, which will be significant for high-speed boats but not significant for lower speed boats and ships. It assumes that the boat hull has been designed properly for the operational speed. Rectangular shaped barges and other oddly shaped boats will have a higher drag force that is predicted by the model.

Example 3.1: The record for a single person rowing 2000 m is 5 min, 36.6 s. The single scull boat used was 27 ft long and had a weight of about 70 lb. The rower weighed about 180 lb. Use the method above to estimate the drag force on the boat and the power required (**Figure 3.2**).

An Olympic athlete in a specially designed boat like this can provide enough power to exceed the hull speed of the boat. The average speed of the boat is calculated as:

$$\text{Speed} = \frac{2000 \text{ m}}{5(60) + 36.6 \text{ s}} = 5.94 \text{ m / s} \tag{3.23}$$

The waterline length of the hull will be approximately 90% of the 27 feet = 24.3 ft = 7.407 m. The hull speed can be calculated from Eq. 3.8 as:

$$\text{Hull Speed} = (0.3985)(9.81)(7.407) = 3.4 \text{ m / s} \tag{3.24}$$

The speed of the scull is greater than 85% of the hull speed, and we know the weight and speed of the boat. Equations 3.11 through 3.14 can be used to estimate the power

FIGURE 3.2   Rowing a single scull.

the rower must provide. The weight of the boat and rower is 250 lb. The speed is 5.94 m/s, which must be converted to knots, and 1 m/s = 1.94384 knots. The speed is 11.55 knots. Going through the calculations yields: $x_D = -4.419$, $x_S = 16.893$, $x_p = 38.204$, Power = 0.759 hp = 566 watt.

In this example, there is no motor, and shaft power is being interpreted as rowing power. The reader should recognize that this is a good way to estimate the power the rower needs, but it is a rough estimate. For this example, the rower would need to provide 566 W for 5 min and 36.6 s. It would take an Olympic athlete to provide that much power. Providing 566 W is a lot of power for a human being, though it is possible. Assuming the efficiency of the oars in the water is comparable to a propeller this is a good estimate of the power the rower must provide.

As a follow-up to this question, we could explore how long it would take a more typical athletic person to row the scull through the 2000 m course. Most young men can train up to providing about 200 W power. Assuming a weight of 250 lb and a power of 200W, Eqs. 3.15 through 3.18 can be used to estimate the speed of the scull. Going through the calculations: $x_D = -4.419$, $x_p = 42.342$, $x_S = 18.962$, Speed = 7.07 knots = 3.636 m/s. It would take a more typical athlete 9 min, 10.9 s to travel the 2000 m course.

The drag force on the scull is equal to the power divided by the speed. For the Olympic athlete the drag force is:

$$\text{Drag Force} = \frac{566\,\text{W}}{5.94\,\text{m/s}} = 95.3\,\text{N} \tag{3.25}$$

Example 3.2: The cruise ship Oasis will carry 6298 passengers and has a reported fuel economy of 14.4 passenger miles per gallon of diesel fuel. (The ship will use a less expensive and dirtier version of diesel fuel when at sea.) The ship is 360 m long and weighs 100,000 metric tons. It is reported to have 60 MW power for propulsion. The cruising speed is 26 mph. Assume that the engines and propeller system is about 25% efficient in converting the chemical energy in the fuel into useful mechanical thrust for the ship.

a. Use the method in the class to estimate the drag force and cruising power for the ship.

b. What is the fuel economy in mpg? In feet per gallon? Assume the fuel has 138,700 BTU/gal.

c. Most of the fuel energy is used to provide electricity for passengers and amenities. Based on 6298 passengers and 14.4 passenger miles per gallon, what fraction of the fuel is used for propulsion? (**Figure 3.3**)

Large ships travel at speeds below 85% of the hull speed with few exceptions, but it is necessary to calculate the ship speed and hull speed to ensure the ship is traveling at a speed below 85% of the hull speed. A speed of 26 mph converts to 11.62 m/s. The waterline length is estimated at 90% of the total length of the ship.

$$\text{Waterline Length L} = (0.9)(360) = 324\,\text{m} \qquad (3.26)$$

$$\text{Hull Speed} = (0.3985)\sqrt{gL} = (0.3985)\sqrt{(9.81)(324)} = 22.46\,\text{m/s} \qquad (3.27)$$

FIGURE 3.3   Cruise ship Oasis.

The ship's speed of 11.62 m/s is less than 85% of the hull speed. Eq. 3.9 and Eq. 3.10 will be used to calculate the drag force and power for the ship. The Froude number is calculated as:

$$Fr = \frac{11.62}{\sqrt{(9.81)(324)}} = 0.20611 \tag{3.28}$$

The mass of the ship is 100,000 metric tons, which is 100×10^6 kg. The drag force and power required are calculated as:

$$\text{Drag Force} = (0.113)(100 \times 10^6)(9.81)(0.20611^2) = 4.709 \times 10^6 \text{ N} \tag{3.29}$$

$$\text{Drag Power} = (4.709 \times 10^6)(11.62) = 54.7 \times 10^6 \text{ W} = 54.7 \text{ MW} \tag{3.30}$$

According to the specifications, the ship has 60 MW of power available for propulsion. The analysis in this chapter yields a reasonable estimate, though probably a high estimate for the drag force and propulsion power at cruising speed. Based on this estimate, we can calculate the fuel consumption and fuel economy for the ship. If the product of engine efficiency and propeller efficiency is 25% as stated in the problem, the fuel consumption can be calculated as:

$$\text{Fuel Usage} = \frac{54.7 \times 10^6 \text{ W}}{0.25} \left(\frac{3.412 \text{ BTU}/\text{h}}{W} \right) \left(\frac{\text{gal}}{138{,}700 \text{ BTU}} \right) = 1300 \text{ gph} \tag{3.31}$$

$$\text{Fuel Economy} = \frac{26 \text{ mph}}{1300 \text{ gph}} = 0.0200 \text{ mpg} = 106 \text{ ft}/\text{gal} \tag{3.32}$$

In the literature, large ships are reported to have fuel economy in the range of 50 ft/gal to 150 ft/gal for propulsion depending on the size and weight of the ship. The Oasis is a cruise ship and must provide electricity for lights, air conditioning, and entertainment of the passengers. The ship also spends long periods docked in a port using no energy for propulsion, but still consuming energy for the passengers. It turns out that propulsion is less than half of the energy consumption for the ship, for a typical cruise ship.

The literature on the Oasis reports that it averages 14.4 passenger mi per gal of fuel consumed. The ship holds 6298 passengers when fully loaded. If we divide the passenger mi per gal by the number of passengers, we can obtain the average fuel economy for the ship.

$$\text{Average Fuel Economy} = \frac{14.4}{6298} = 0.002286 \text{ mpg} = 12.07 \text{ ft}/\text{gal} \tag{3.33}$$

In traveling a mi, the ship would use (5280/106) = 50 gals of fuel for propulsion. The total fuel consumption, on average, for one mile traveled would be (5280/12.07) = 437 gals because of the fuel being used for things other than propulsion.

A large fraction of the energy is being consumed while the ship is in port. Overall, the ship uses about 11.4% of the fuel for propulsion and 88.6% for other things.

For a 300-mi journey at 26 mph, the ship would use (300 mi) (50 gal/mi) = 15,000 gals of fuel for propulsion. When at sea, ships can use a diesel fuel that has a higher sulfur content than is allowed in port. The "dirtier" fuel is less expensive and would cost about $1.50 per gallon. The cost of propulsion for the 300 mi trip would be $22,500. Dividing by the 6298 passengers yields a cost of $3.57 per passenger. The cost of propulsion is an insignificant part of the cost of taking a cruise.

If we factor in an average of 437 gal/mi, the total fuel cost is $196,650.00. Dividing by the 6298 passengers on board the cost is $31.22 per passenger. That is still a minor part of the cost of taking the cruise.

Example 3.3: A Formula 240 Bowrider is powered by a 320 hp gasoline engine. The boat is 31 ft long and has an empty weight of 5000 lb. Cruising speed for the boat is 32 mph and is rated to use 11 gph of fuel at that speed. Assume that the boat is carrying 1500 lb of people and gear.

   a. Calculate the shaft power to move it along at 32 mph.

   b. Compare the calculations to the fuel consumption rate of 11 gph. Assume that the engine and propeller are 20% efficient in converting the energy in the fuel to useful thrust. Assume the gasoline has 121,000 BTU per gallon (**Figure 3.4**).

FIGURE 3.4   Formula 240 Bowrider.

© Shutterstock

The speed of 32 mph is converted to 14.30 m/s. The waterline length is estimated as 90% of the total 21 ft length of the boat, which is 8.504 m. The hull speed is calculated as 3.64 m/s, which means the boat speed is greater than 85% of the hull speed. The method associated with Figure 3.1 will be used to analyze the boat. The total weight of the boat is 6500 lb and the speed is 14.30 m/s = 27.80 knots. Eqs. 3.11 through 3.14 will be used to estimate the power required. If the weight of the boat is 6500 lb, and the speed is 27.80 knots, calculations are: $x_D = 8.434$, $x_S = 13.19$, $x_P = 17.95$, Shaft Power = 124 hp.

The calculations show that the shaft power is 124 hp. The engines are rated at 320 hp, which is significantly higher than the 124 hp calculated to power the boat. There will be power lost in the transmission, but the engines have enough power for the boat to cruise at 32 mph. The propellers attached to the engines will be less than 100% efficient, so the actual thrust power for the boat will be less than 124 hp.

There is another way to estimate the power requirement of the engines from the information given. The information says the boat will use 11 gph gasoline when cruising at 32 mph. If we assume the engine has a thermal efficiency of 28% and the propeller has an efficiency of 70%, and that the fuel has an energy content of 121,000 BTU per gal:

$$\text{Power} = \left(11\,\text{gph}\right)\left(121{,}000\,\frac{\text{BTU}}{\text{gal}}\right)(0.28)(0.70)\left(\frac{\text{hp}}{2545\dfrac{\text{BTU}}{\text{h}}}\right) = 102.5\,\text{hp} \quad (3.34)$$

The answers of 124 hp and 102.5 hp are in reasonably good agreement when you consider the assumptions made in the analysis. If we had assumed the propeller was 85% efficient instead of 70% efficient, the answers would be in exact agreement.

The analysis in this section is reasonably good for large ships traveling well below the hull speed. Boats are very difficult to analyze because the boat rises and planes on the water as it speeds up. The analysis is complex and the equations given in this section should be regarded as only rough approximations.

3.1.2 Propellers

Propellers accelerate the flow of the fluid. The speed of the air approaching the propeller is less than the speed of air exiting the propeller. Conservation of mass requires that the intake area be larger than the exhaust area. $V_1 A_1 = V_2 A_2$, so if $V_2 > V_1$, then $A_1 > A_2$. V_1 is the speed of the vehicle.

Figure 3.5 illustrates how the intake area must be greater than the exhaust area because the exhaust velocity is greater than the intake velocity. There is no physical boundary around the propeller. The boundary illustrates the control volume and streamlines.

The derivation by Froude starts with the conservation of mass. It is assumed that the area at the propeller is πR^2, where R is the radius of the propeller. V_1 is the intake velocity on the left side of the control volume, V is

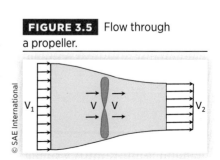

FIGURE 3.5 Flow through a propeller.

© SAE International

the velocity at the propeller, and V_2 is the velocity as the flow exits the control volume. Conservation of mass says that the mass flow rate must be the same at all three points, and mass flow is the product of the density of air, the area, and the velocity of the flow.

$$\rho V_1 A_1 = \rho V_2 A_2 = \rho V \left(\pi R^2 \right) \tag{3.35}$$

For the conservation of momentum equation, there is a thrust force F acting on the propeller. The thrust force F is what propels the airplane or ship. Applying the conservation of momentum principle to the fluid it follows:

$$F = -V_1 \, \rho \, V_1 \, A_1 + V_2 \, \rho \, V_2 \, A_2 \tag{3.36}$$

Substituting the conservation of mass equation into the conservation of momentum equation yields the following equation for the thrust force F:

$$F = \rho V \left(\pi R^2 \right) \left(V_2 - V_1 \right) \tag{3.37}$$

At this point in the derivation, an assumption or approximation must be made to develop a useful formula. Froude assumed that the velocity immediately before and after the propeller was equal and equal to the flow velocity through the propeller. This leads to the equation below, though an equivalent way to do the derivation is to simply assume the equation below is correct:

$$V = \frac{V_1 + V_2}{2} \tag{3.38}$$

Equation 3.37 can be substituted into Eq. 3.36 to yield the thrust equation. V_1 is the velocity of the airplane or ship in the fluid. V_2 is the velocity of the fluid exiting the control volume relative to the airplane or ship. V_2 is not an absolute velocity. It is the velocity relative to the vehicle (airplane or ship).

$$F = \frac{1}{2} \rho \left(\pi R^2 \right) \left(V_2^2 - V_1^2 \right) \tag{3.39}$$

Equation 3.39 can be used to calculate the thrust force the propeller provides to the vehicle. The power provided to the vehicle is the thrust force multiplied by the velocity, which is V_1 in this case.

$$\text{Power Provided to Vehicle} = F \, V_1 \tag{3.40}$$

The power being consumed by the propeller is the thrust force F multiplied by the velocity of the fluid through the propeller V. Since V is always greater than V_1, the power required by the propeller is always more than the power provided to the vehicle. A thrust efficiency (e_{prop}) for the propeller is defined as the ratio of the power provided to the vehicle divided by the power provided to the propeller.

$$e_{prop} = \frac{F \, V_1}{F \dfrac{V_1 + V_2}{2}} = \frac{2 \, V_1}{V_1 + V_2} \tag{3.41}$$

The efficiency definition in Eq. 3.41 is useful once the vehicle gets to cruising speed. It gives a way to calculate the efficiency of the propeller discussed in previous sections.

Unfortunately, the definition says that the propeller efficiency is zero when the velocity of the vehicle is zero, such as when an airplane is beginning it's run down the runway on takeoff. Mathematically, the efficiency is zero when the velocity of the vehicle is zero, but the definition does not fully capture what we mean by efficiency in that case. The propeller provides the thrust necessary to accelerate the vehicle to speed and is very important in making the vehicle functional. To say the efficiency is zero when the vehicle velocity is zero does not fully capture what is happening with the vehicle. But this is the definition that is used for propeller efficiency.

Example 3.4: An 18-ft long, 2500-lb weight bass boat is traveling at 37 mph. Use the method from the previous section to estimate the drag force on the boat. The transmission is 85% efficient in transmitting the engine power to the propeller. The propeller has a diameter of 14 inches. Find the thrust power that the engine must provide to push the boat along at 37 mph. Assume the density of water is 1000 kg/m³ (**Figure 3.6**).

The 37 mph speed of the bass boat is greater than 85% of the hull speed. The weight and speed of the boat are given and we want to calculate the power required. Equations 3.11-3.14 are used to estimate the shaft power required. The speed of 37 mph is converted to 32.15 knots. If the weight of the boat is 2500 lb, and the speed is 32.15 knots, calculations are: $x_D = 4.664$, $x_S = 12.58$, $x_P = 20.50$, Shaft Power = 65.27 hp.

The thrust power provided to the boat is less than the shaft horsepower because the propeller will not be 100% efficient. The engine will need to provide 65.27 hp to the propeller, but the thrust power provided to the boat will be less than 65.27 hp.

FIGURE 3.6 Bass boat.

© Shutterstock

The problem is solved below using metric units. The speed $V_1 = 37$ mph is converted to 16.5369 m/s. The propeller radius of 7 is converted to 0.1778 m. The shaft power is equal to the drag force F on the boat multiplied by the velocity of the fluid at the propeller:

$$\left(65.27\,\text{hp}\right)\left(\frac{745.7\,\text{W}}{\text{hp}}\right) = F\left(\frac{16.5369 + V_2}{2}\right) \tag{3.42}$$

Equation 3.39 is used to calculate the drag force F:

$$F = \frac{1}{2}\left(1000\right)\left(\pi\,0.1778^2\right)\left(V_2^2 - 16.5369^2\right) \tag{3.43}$$

Equations 3.42 and 3.43 are solved to find $V_2 = 18.165$ m/s and F = 2805 N. The thrust power that the propeller is providing to the boat is the drag force F multiplied by the speed V_1:

$$\text{Thrust Power} = \left(2805\,\text{N}\right)\left(16.5369\,\text{m/s}\right) = 46{,}389\,\text{W} = 62.2\,\text{hp} \tag{3.44}$$

Example 3.5: Suppose we are planning to use a pair of 7,000 hp diesel engines to power a ship that will cruise at 25 mph, and we want to estimate the size of the propeller that will be required. A higher efficiency propeller will allow the ship to burn less fuel, which reduces the operating cost. We estimate that it will take 500 kN thrust to push the ship through the water. Assume that the engines/drives are 32% efficient and the fuel cost is $1.25 per gallon. The ship is to cruise from Los Angeles to Hawaii (2500 mi), from Hawaii to Midway (1500 mi), and from Midway to Tokyo (2600 mi), refueling at each stop. The density of seawater is 1025 kg/m³.

a. Size the propeller assuming a 75% efficiency at the cruising speed. Calculate the cost of the fuel for a one-way trip.

b. Explore the cost difference using a 70% efficient propeller and an 80% efficient propeller (**Figure 3.7**).

The speed of the ship is $V_1 = 25$ mph. Eq. 3.41 is used to find the velocity V_2 of the water relative to the ship:

$$0.75 = \frac{2(25)}{25 + V_2} \quad V_2 = 41.67\,\text{mph} \tag{3.45}$$

It will be convenient to work the problem in the metric system, so the velocities need to be converted to units of m/s. $V_1 = 25$ mph = 11.17 m/s, and $V_2 = 41.57$ mph = 18.62 m/s. Ships normally refer to the propeller as the screw. The radius of the screw can be calculated using the thrust formula (Eq. 3.39):

$$500{,}000\,\text{N} = \frac{1}{2}\left(1025\right)\pi R^2\left(18.62^2 - 11.17^2\right) \tag{3.46}$$

FIGURE 3.7 Small ship.

© Shutterstock

$$R = 0.83644\,\text{m}, D = 1.673\,\text{m} \tag{3.47}$$

The power required on the shafts driving the propellers is equal to the drag force multiplied by the average velocity through the screw:

$$\text{Shaft Power} = \left(500{,}000\right)\left(\frac{11.17 + 18.62}{2}\right) = 7.45\,\text{MW} = 9987\,\text{hp} \tag{3.48}$$

The power would be divided between two 7000 hp engines, so the engines will have enough power to push the ship along. The total time required to travel from Los Angeles to Tokyo can be calculated as:

$$\text{Time} = \frac{2500\,\text{mi} + 1500\,\text{mi} + 2600\,\text{mi}}{26\,\text{mph}} = 264\,\text{h} \tag{3.49}$$

The energy required is the power multiplied by the time. This should be converted to units of BTU to calculate the number of gallons of fuel required.

$$\text{Energy} = \left(7.45 \times 10^{6}\,\text{W}\right)\left(264\,\text{h}\right)\left(\frac{3.412\,\text{BTU}}{\text{W-h}}\right) = 6.708 \times 10^{9}\,\text{BTU} \tag{3.50}$$

The engine/drive system is 32% efficient in getting the power to the shafts that power the screws. Diesel fuel has 138,700 BTU/gal, and the cost is assumed to be $1.25 per gal. The number of gallons and cost is calculated below:

$$\text{Fuel} = \frac{6.708 \times 10^9 \text{ BTU}}{(0.32)(138,700 \text{ BTU / gal})} = 151,147 \text{ gals} \tag{3.51}$$

$$\text{Cost} = (151,147 \text{ gals})(\$1.25 / \text{gal}) = \$188,934.00 \tag{3.52}$$

These calculations were done assuming a 75% efficient propeller. The same process can be repeated for the 70% efficient propeller. The V_2 velocity is 46.43 mph = 20.75 m/s. The diameter of the propeller required is 1.504 m. Power required on the shafts driving the screws is 7.98×10^6 W. Making the trip from Los Angeles to Tokyo would require 161,953 gals of fuel for $202,441.00.

The same process can be repeated for the 80% efficient propeller. The V_2 velocity is 37.5 mph = 16.76 m/s. The diameter of the propeller required is 1.995 m. Power required on the shafts driving the screws is 6.9825×10^6 W. Making the trip from Los Angeles to Tokyo would require 141,709 gals of fuel for $177,136.00.

The analysis shows the advantage of using a more efficient propeller (screw). Generally, the more efficient propellers will be larger in diameter than the less efficient propellers. The fuel savings and cost savings that go with using a more efficient propeller are very significant. The design of a propeller is beyond the scope of the book, but the readers should see that using a larger and more efficient propeller is almost always worth the extra cost.

3.2 Homework

1. A small aluminum fishing boat is to be powered by a 5 hp gasoline motor. When cruising across the lake the motor is at partial throttle such that it is producing 3 hp. The transmission and propeller are 75% efficient in converting the engine power into thrust power for the boat, so there is 2.25 hp of actual thrust power on the boat. The total weight of the boat, two fishermen, and their gear is 550 lb. The boat is 14 feet long, and the waterline length is 90% of the boat length. Assume that the engine has a thermal efficiency of 22% and the gasoline has an energy density of 121,000 BTU/gallon.

    a. Calculate the hull speed for the boat. Express your answer in ft/s.

    b. Assume that the boat will be cruising at more than 85% hull speed and calculate the cruising speed for the boat when the engine is producing 3 hp. Verify that the boat is traveling at more than 85% hull speed.

c. With a 3-gal tank of fuel, how far can the boat travel at the cruising speed? (Answer 200 mi) (**Figure 3.8**)

FIGURE 3.8   Bass Boat.

© Shutterstock

2. A large ship has a total weight of 54,500 metric tons and cruises at 28 mph on the ocean. The waterline length on the ship is 325 m. The density of seawater is 1025 kg/m³.

a. Calculate the drag force and required power for the ship at cruising speed.

b. Assume that the ship is powered by 3 propellers that are 5.5 m in diameter. Calculate the required engine horsepower for the propeller system. (Answer 41.45 Mwatts)

c. Assume that the diesel engines are 30% efficient and the fuel has 138,700 BTU per gallon. Calculate the fuel economy in mpg and feet per gallon. How much fuel is required to make a 300-mi trip? (Answer 36,400 gals) (**Figure 3.9**)

FIGURE 3.9 Large ship.

© Shutterstock

3. Assume that the density of seawater is 1025 kg/m³. A 15,000-lb fishing boat powered by twin 120 hp motors is traveling at 18 knots. The static waterline length of the boat is 30 feet. The propellers have a diameter of 18 inches. (Knot is a speed of 1 nautical mile per hour. A nautical mile is equal to 1.151 mi.)

a. Find the drag force on the boat and the efficiency of the propeller being used. Assume that the drag force is shared equally by the two propellers. (Answer: Drag force = 8089 N, V2 = 11.57 m/s, efficiency = 88.9%)

b. Assume that the engines burn gasoline that has an energy density of 125,000 BTU per gallon, and that the engines are 22% efficient in converting the energy in the fuel into torque on the propeller shaft. The cost of fuel is $3.18 per gallon at the marina. How many gallons per hour does the boat use while cruising at 18 knots? What is the fuel cost to power the boat at this speed for 3 h? (10.45 gals per hour, $99.74) (**Figure 3.10**)

FIGURE 3.10 Fishing boat.

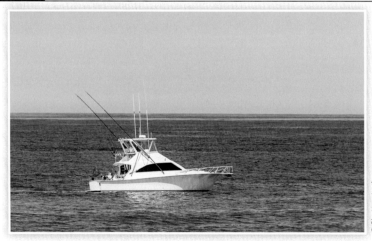

© Shutterstock

4. This problem builds on the fishing boat in problem 3. Please reference the information in problem 4. The owner finds a less expensive propeller for the engines that has a diameter of 16 inches and considers switching to the less expensive propeller. Assume that the speed (18 knots) and drag force on the boat are the same for both propeller choices.

   a. Calculate the exit velocity V_2 required for the 16-inch diameter propeller (Use the thrust formula). ($V_2 = 12.1$ m/s)

   b. Calculate the efficiency of the 16-inch diameter propeller.

   c. Calculate the power required on the propeller shaft. Is this more or less or the same as in problem 3? What is the fuel cost to power the boat for 3 h at 18 knots? Compare the cost to the cost in problem 4. (115.9 hp, $102.33)

Reference

1.   https://en.wikipedia.org/wiki/Froude_number

4

Airplanes, Jets, and Rockets

4.1 Drag Force for Airplanes and Jets

The drag force acting on airplanes and jets is due to the aerodynamic drag. Propulsion comes from the propellers on airplanes and from the jet engine on jets. For steady-state flight, the drag force on the airplane must equal the propulsion force from the propellers or jet engines.

Airplanes and jets must produce enough lift to offset their weight. For steady-state flight, the lift force on the aircraft must equal the weight. The wings of the aircraft must be tilted upward slightly to generate the required lift force. Tilting the wings increases the aerodynamic drag force. The increase in aerodynamic drag due to tilting the wings and generating the required lift is known as induced drag.

The drag force acting on the aircraft is broken into two parts, namely, basic drag and induced drag. The aircraft is designed for a specific cruising speed to minimize the total drag force on the aircraft at the cruising speed. This allows the aircraft to get the best fuel economy possible. Jets burn a lot of fuel. The weight of the fuel is a significant portion of the total weight of the jet. The cost of the fuel is a significant part of the operating cost for a jet. Optimizing for fuel efficiency at the cruising speed is an important design consideration.

Glide Ratio and Lift to Drag Ratio: If the power is cut, an airplane or jet will glide at a downward angle θ, as illustrated in **Figure 4.1** [1]:

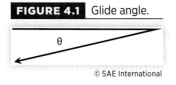

FIGURE 4.1 Glide angle.

The downward vertical component velocity is V sin(θ). Assuming a constant speed gliding, the gravitational power (mg V sin(θ)) must equal the power to overcome aerodynamic drag. The conservation of energy principle is used to formulate an equation for the drag force on the aircraft:

$$mg\,V\sin(\theta) = (\text{Drag Force})V; \text{Drag Force} = mg\sin(\theta) \qquad (4.1)$$

If we know the glide angle θ of the airplane or jet in flight, the equation above can be used to find the drag force on the airplane or jet. The glide angle θ varies from aircraft to aircraft and varies with the speed of the aircraft. The glide angle will be smallest at the designed cruising speed for the aircraft.

Rather than reporting a glide angle, aircraft manufacturers report the Glide Ratio for the aircraft, which is the cosecant of the glide angle. A glide ratio of ten means that the aircraft travels 10 m for every 1 m it drops in altitude. The glide angle for this case is arcsin (1/10) = 5.74°.

The Glide Ratio is equal to the Lift to Drag Ratio. An aircraft with a Glide Ratio of ten will generate 10 lb of lift force for every 1 lb of drag force.

a. Small single-engine airplanes with fixed landing gear typically have a Glide Ratio of about ten.

b. Airplanes with retractable landing gear typically have a glide ratio of about 15.

c. Older jets like the Boeing 747 typically have a Glide Ratio of about 16.

d. Newer jets typically have a glide ratio of about 20.

These glide ratios are considered when the aircraft is operating at the cruising speed. The glide ratio will be lower at other speeds. Be careful looking up the Glide Ratio numbers on the internet. The values posted are often incorrect. This chapter covers how to estimate the fuel consumption and fuel economy of airplanes and jets operating at their designed cruising speed. An important part of the analysis is to make a good estimate of the drag force on the aircraft, and this requires a good estimate of the glide ratio. Use the guidance given in the above paragraph to estimate the glide ratio for the airplane or jet. The thrust power required to keep the aircraft in flight is calculated from Eq. 4.2 and Eq. 4.3:

$$\text{Drag Force} = \frac{\text{Aircraft Weight}}{\text{Glide Ratio}} \qquad (4.2)$$

$$\text{Thrust Power} = (\text{Drag Force})(V) \qquad (4.3)$$

4.2 Fuel Economy of Airplanes

Equation 4.2 and Eq. 4.3 are used to estimate the drag force and thrust power required for airplanes and jets. Airplanes are driven using propellers and jets use jet engines. The examples below illustrate how to use the propeller formula to estimate the fuel economy of airplanes.

Example 4.1: A single-engine Cessna 350 Corvalis airplane with fixed landing gear has a mass of 1300 kg and cruises at 200 mph. Estimate the drag force on the airplane and the cruising power (**Figure 4.2**).

Since this is a small airplane with fixed landing gear, the glide ratio will be approximately 10. Drag force and thrust power are calculated as follows:

$$\text{Drag Force} = \frac{(1350)(9.81)}{10} = 1275 \text{ N} \tag{4.4}$$

$$\text{Thrust Power} = (1275)(200)\left(\frac{1609}{3600}\right) = 114.0 \text{ kW} = 152 \text{ hp} \tag{4.5}$$

The propeller will need to generate 1275 N of thrust to keep the plane in flight. The propeller will be less than 100% efficient, so the engine will need to provide more than 152 hp to the propeller to get the required 1275 N thrust.

The next step in estimating fuel consumption is to calculate the power the engine must provide. Assume that a 6 ft diameter propeller is used to power the Columbia 350 airplane and estimate the engine power required. Assume the density of air is 1.2 kg/m³. The speed of the airplane is 200 mph, which is converted to 89.39 m/s. The radius of the propeller is 3 ft, which is converted to 0.9144 m.

The thrust force from the propeller must be 1275 N. The thrust force formula (Eq. 3.39) can be used to find the required value for V_2, which is the velocity of the air leaving the propeller with respect to the velocity of the airplane.

$$1275 \text{ N} = \frac{1}{2}(1.2)\pi\left(0.9144^2\right)\left(V_2^2 - 89.39^2\right); V_2 = 93.81 \text{ m/s} \tag{4.6}$$

FIGURE 4.2 Cessna 350 Corvalis.

Courtesy of Roberto Bianchi

The engine power is the thrust force multiplied by the average velocity through the propeller.

$$\text{Engine Power} = (1275)\left(\frac{89.39 + 93.81}{2}\right) = 116.8\,\text{kW} = 156.6\,\text{hp} \tag{4.7}$$

The 6 ft diameter propeller has a high efficiency at the cruising speed for this airplane. The efficiency formula (Eq. 3.41) yields:

$$\text{Propeller Efficiency} = \frac{2(89.39)}{89.39 + 93.81} = 97.9\% \tag{4.8}$$

The propeller formula used in the analysis are a bit idealistic, and the propeller would probably not be quite this efficient, but this method is a good first estimate for power and efficiency of the propeller. Once the required horsepower of the engine is known, the fuel consumption and fuel economy is estimated using the same method as for wheeled vehicles. Assume that the engine is 25% efficient in converting the energy in the fuel into power, and that the fuel has 120,200 BTU/gallon. Calculate the fuel economy of the airplane in mpg.

$$\text{Fuel Power} = \frac{116,790\,\text{W}}{0.25} = 467,160\,\text{W} \tag{4.9}$$

$$(467,160\,\text{W})\left(\frac{3.412\,\text{BTU}}{\text{W} - \text{h}}\right)\left(\frac{\text{gal}}{120,200\,\text{BTU}}\right) = 13.26\,\text{gph} \tag{4.10}$$

$$\text{Fuel Economy} = \frac{200\,\text{mph}}{13.26\,\text{gph}} = 15.1\,\text{mpg} \tag{4.11}$$

The Columbia 350 has a fuel tank of 102 gals and the manufacturer reports a range of 1500 mi. The calculations in this example (15.1 mpg) (102 gals) = 1540 mi is in good agreement with actual performance of the airplane.

Example 4.2: A Bombardier Dash 8 passenger plane has a mass of 16,466 kg and cruises at 332 mph. It has retractable landing gear, so the glide ratio will be approximately 15. Assume that the propellers are 8 ft in diameter and that the density of air is 1.2 kg/m³. The aircraft has an 835-gal tank and the manufacturer says it has a range of 1300 mi. Assume the engine has a thermal efficiency of 25%.

a. Estimate the drag force on the airplane when cruising.

b. Use the propeller formula to estimate the engine power required for cruising.

c. Assume that the engine has a thermal efficiency of 25% and that it burns aviation fuel which has an energy density of 120,200 BTU/gallon. Calculate the fuel economy of the airplane in mpg.

d. Based on the 835-gal tank and fuel economy when cruising, estimate the range of the airplane. How does this compare with the manufacturer's estimate of 1300 mi? (**Figure 4.3**)

FIGURE 4.3   Bombardier Dash 8 airplane.

The solution to the problem is worked out using the metric system below.

$$\text{Drag Force} = \left(\frac{(16,466\,\text{kg})(9.81\,\text{m/s})}{15} \right) = 10,756\,\text{N} \qquad (4.12)$$

The velocity of the airplane V_1 is 332 mph, which is converted to 148.4 m/s. The radius of the propellers is 4 ft, which is converted to 1.219 m. The thrust force formula (Eq. 3.39) is used to find the velocity of the air behind the propellers relative to the airplane V_2.

$$10,756\,\text{N} = \frac{1}{2}(1.2)\pi\left(1.219^2\right)\left(V_2^2 - 148.4^2\right)(2\,\text{propellers}); V_2 = 154.7\,\text{m/s} \quad (4.13)$$

$$\text{Engine Power} = (10,756)\left(\frac{148.4 + 154.7}{2} \right) = 1.630 \times 10^6\,\text{W} \qquad (4.14)$$

$$\text{Fuel Power} = \left(\frac{1.630 \times 10^6\,\text{W}}{0.25} \right)\left(\frac{3.412\,\text{BTU}}{\text{W}-\text{h}} \right)\left(\frac{\text{gal}}{120,200\,\text{BTU}} \right) = 185.1\,\text{gph} \quad (4.15)$$

$$\text{Fuel Economy} = \frac{332\,\text{mph}}{185.1\,\text{gph}} = 1.79\,\text{mpg} \qquad (4.16)$$

The fuel tank holds 835 gals according to the manufacturer, so the range can be calculated as:

$$\text{Range} = (835\,\text{gals})(1.79\,\text{mpg}) = 1498\,\text{mi} \qquad (4.17)$$

The manufacturer says the range is 1300 mi, which is less than calculated. The calculations above assume the miles are spent cruising at the operating speed. Allowing for takeoff and landing will require extra fuel and reduce the range, and the plane will need to have a reserve left in the tank when it lands for safety reasons. All things considered, the method above makes a good estimate for the fuel economy of the aircraft in flight.

In addition to being able to power the engine during flight, the engines and propellers must be able to provide enough thrust to accelerate the airplane up to speed on the runway. The maximum power from the engines will be used during takeoff. As the airplane accelerates down the runway there is a thrust force from the engines and propellers and an aerodynamic drag force on the airplane that increases with the square of the speed. Assuming the airplane has a mass m and acceleration a, Newton's second law can be written as:

$$\text{Thrust Force} - \text{Aerodynamic Drag Force} = ma \tag{4.18}$$

Thrust force and aerodynamic drag are functions of velocity of the airplane. Acceleration is the derivative of velocity with time. It is possible to set up a first-order differential equation with velocity as the unknown, but the nonlinear nature of the functions for thrust force and aerodynamic drag makes it difficult to find an analytic solution. An approximate solution can be developed using a spreadsheet approach. The solution method is illustrated best with an example.

Example 4.3: A turboprop airplane must accelerate to 160 mph on the runway to takeoff. The plane has a total mass of 60,000 kg. The two engines and 3 m diameter propellers push the air backward at a speed of $V_2 = 500$ mph relative to the aircraft when at full throttle. Assume that the aerodynamic drag force on the airplane varies with the square of speed, Aero Drag $= 5.5\ V^2$, where V is in m/s and the force is in N. Develop a spreadsheet model for acceleration of the plane down the runway at full throttle. Assume the density of air is 1.2 kg/m³.

Develop the spreadsheet model for the plane going down the runway with a time increment of 0.1 s. The conclusion should be that it takes about 10.8 s for the airplane to takeoff, a distance of about 400 m down the runway (**Figure 4.4**).

Aerodynamic drag is proportional to the square of the velocity of an object. When we include aerodynamic drag in the model, we always end up with a nonlinear differential equation to solve for the velocity and position of the object as a function of time. A schematic of the forces acting on the aircraft during takeoff is shown below (**Figure 4.5**):

Recognizing that V_2 is given as 500 mph = 223.5 m/s and that the acceleration is a $= dV_1/dt$:

$$\frac{\rho \pi R^2}{2}\left(223.5^2 - V_1^2\right)(2) - (5.5)V_1^2 = m\frac{dV_1}{dt} \tag{4.19}$$

FIGURE 4.4 Turboprop airplane.

© Shutterstock

FIGURE 4.5 Thrust and drag forces on airplane.

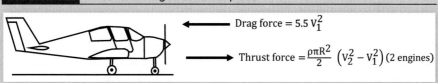

Drag force $= 5.5\,v_1^2$

Thrust force $= \frac{\rho \pi R^2}{2}\left(v_2^2 - v_1^2\right)$ (2 engines)

© SAE International

In this case, it is possible to separate variables and integrate to get V_1 as an analytical function of time, but the result is difficult to work with. For this set of parameters, the analytical solution is:

$$V_1 = \left(174.078\,\mathrm{m/s}\right)\left(\frac{e^{\frac{t}{12.3253}} - 1}{e^{\frac{t}{12.3253}} + 1}\right) \qquad (4.20)$$

As the analysis becomes more complex it becomes impossible to find an analytical solution to the differential equation. We will develop an approximate solution using excel. In the first column increment the time in 0.1 s to 30 s. The approximation used is that the acceleration is constant over a 0.1-s increment in time.

It is difficult to present spreadsheet development in book form. The equations for the subsequent columns in the spreadsheet below are listed. The reader should work on the spreadsheet until it matches the first several lines printed:

a. For the first line in the spreadsheet V1 is zero. In subsequent lines the velocity is acceleration multiplied by the time increment plus the velocity at the beginning of the time step. That is, for line i+1, $V1_{i+1} = V1_i + a_i(0.1)$. The approximation used in this model is that the acceleration is constant over the time step. The units for V1 in this column are m/s.

b. In the next column the speed V1 is converted to mph by multiplying by 3600/1609.

c. In the next column V2 is a constant value for this model. V2 = 500(1609/3600) to convert it into m/s.

d. The thrust force is calculated from Eq. 3.39 using $\rho = 1.2$ kg/m³, R = 1.5 m, and V2 and V1 are from the fourth and second columns in the spreadsheet. There are two propellers for this plane, so there is a factor of 2 on the thrust formula.

e. The drag force is from the equation in the figure using V1 from column 2.

f. The net force is the thrust force minus the drag force.

g. The acceleration is the net force divided by the mass of 60,000 kg.

h. For the first line in the spreadsheet, the distance the airplane has traveled on the runway is zero. In subsequent lines, the distance is the average velocity during the time step multiplied by the time increment, plus the distance at the beginning of the time step. With the assumption of constant acceleration during the time step, the average velocity is the average of the initial and final velocities during the time step. For line i+1, $d_{i+1} = d_i + (V1_i + V1_{i+1})(0.1)/2$.

i. Propeller efficiency is 2V1/(V1 + V2).

j. The altitude column is included for the next part of the problem where we will model the airplane taking off. For this part of the problem the altitude is zero since the airplane is on the runway.

The first few lines of the spreadsheet are shown in Table 4.1 below (**Table 4.1**):

Follow the first column down to 10.8 s and the speed in the third column is V1 = 160 mph. At this point the plane has traveled about 400 m down the runway. The reader may want to compare the analytical solution (Eq. 4.20) to the approximate solution. At 10.8 s, the analytical solution yields a speed $V_1 = 71.73538$ m/s and the approximate solution yields $V_1 = 71.73618$ m/s. This comparison is included to help convince the reader that the spreadsheet approach yields an accurate solution to the problem. At 160 mph, the airplane takes off and gravity becomes part of the analysis. Gravity can be included easily in the spreadsheet model, but it is more difficult to include it in the analytical solution.

TABLE 4.1 Spreadsheet model example.

s	m/s	mph	m/s	N	N	N	m/s²	m	propeller	m	mph
Time	V1	V1	V2	Thrust	Aero Drag	Net Force	Acceleration	Distance	efficiency	Altitude	Speed
0	0	0	223.4722	423604.7	0	423604.7	7.06007772	0	0	0	0
0.1	0.706008	1.579632	223.4722	423600.4	2.741458	423597.7	7.05996156	0.070601	0.006299	0	1.579632
0.2	1.412004	3.159238	223.4722	423587.8	10.96565	423576.8	7.0596131	0.211801	0.012558	0	3.159238
0.3	2.117965	4.738766	223.4722	423566.6	24.67177	423541.9	7.05903236	0.423598	0.018777	0	4.738766
0.4	2.823868	6.318164	223.4722	423537	43.85828	423493.2	7.05821942	0.705985	0.024957	0	6.318164
0.5	3.52969	7.897381	223.4722	423499	68.52293	423430.5	7.05717436	1.058954	0.031098	0	7.897381
0.6	4.235408	9.476363	223.4722	423452.5	98.66274	423353.8	7.05589732	1.482494	0.0372	0	9.476363
0.7	4.940998	11.05506	223.4722	423397.6	134.274	423263.3	7.05438845	1.976594	0.043264	0	11.05506
0.8	5.646436	12.63342	223.4722	423334.2	175.3523	423158.9	7.05264793	2.541238	0.049288	0	12.63342
0.9	6.351701	14.21139	223.4722	423262.5	221.8926	423040.6	7.05067599	3.176408	0.055274	0	14.21139
1	7.056769	15.78892	223.4722	423182.3	273.8889	422908.4	7.04847288	3.882085	0.061222	0	15.78892

Example 4.4: Assume that once the airplane in the previous problem reaches 160 mph it takes off and climbs at a 10% grade. Because the plane is climbing, there is an additional gravitation force opposing the motion of 10% of the airplane weight. Include the gravitational force and plot the altitude of the plane with time.

To modify the spreadsheet to account for takeoff, go into the net force column at 10.8 s and subtract 10% of the weight of the airplane from the net force to account for the plane traveling upward against gravity. The altitude is estimated assuming that the acceleration is approximately constant for the time step. The change in altitude is 10% of the average velocity multiplied by the time step. For line i+1, altitude$_{i+1}$ = altitude + (0.10) (V1$_i$ + V1$_{i+1}$) (0.1)/2. The changes in the spreadsheet must be copied down so that subsequent lines are changed. The first few lines in the transition area of the spreadsheet are shown below (**Table 4.2**).

There are many plots that can be obtained from the data. Plotting the airplane speed (V1), thrust force, aero drag force, net force, acceleration, distance on the runway, or propeller efficiency versus time are all interesting charts. I have chosen to include the plot of altitude versus time to illustrate the airplane going down the runway and taking off, the first 30 s. The airplane takes off and climbs to 208 m in the first 30 s (**Figure 4.6**).

TABLE 4.2 Spreadsheet transition to takeoff.

s	m/s	mph	m/s	N	N	N	m/s²	m	propeller	m	mph
					Aero	Net					
Time	V1	V1	V2	Thrust	Drag	Force	Acceleration	Distance	efficiency	Altitude	Speed
10.5	69.26585	154.9764	223.4722	382908.6	26387.67	350635	5.84391625	378.4191	0.473227	0	154.9764
10.6	69.85024	156.2839	223.4722	382219	26834.81	349498.2	5.82497063	385.4042	0.476269	0	156.2839
10.7	70.43274	157.5872	223.4722	381525.9	27284.24	348355.7	5.80592804	392.4474	0.479289	0	157.5872
10.8	71.01333	158.8863	223.4722	380829.3	27735.91	347207.4	5.78679033	399.5488	0.482287	0	158.8863
10.9	71.59201	160.181	223.4722	380129.3	28189.79	346053.6	5.76755938	406.708	0.485264	0.71592	160.181
11	72.16876	161.4714	223.4722	379426	28645.82	344894.2	5.74823703	413.9248	0.488219	1.43760	161.4714
11.1	72.74359	162.7576	223.4722	378719.5	29103.96	343729.5	5.72882516	421.1992	0.491153	2.16504	162.7576
11.2	73.31647	164.0393	223.4722	378009.7	29564.18	342559.5	5.70932561	428.5309	0.494065	2.89820	164.0393
11.3	73.8874	165.3168	223.4722	377296.8	30026.42	341384.4	5.68974026	435.9196	0.496957	3.63708	165.3168
11.4	74.45638	166.5898	223.4722	376580.9	30490.64	340204.3	5.67007095	443.3652	0.499827	4.38164	166.5898
11.5	75.02338	167.8584	223.4722	375862	30956.8	339019.2	5.65031953	450.8676	0.502677	5.13188	167.8584

FIGURE 4.6   Plot of altitude versus time for the first 30 s.

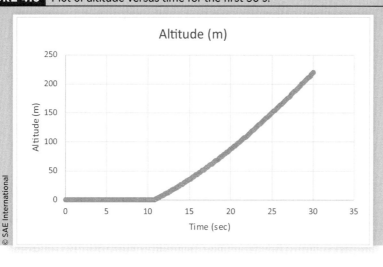

4.3 Fuel Economy of Turbojet and Turbofan Aircraft

For turbojet and turbofan engines we will need to analyze the mass flow rate through the engine. Air flows into the engine, fuel is added and burned, and the combustion products flow out of the engine. There is more mass flowing out of the engine than flowing in because of the fuel added. For the turbojet engine all the air flows through the combustion chamber. For turbofan engines some of the air flows around the combustion chamber and is added and heated with the exhaust gasses. Having a high air to fuel ratio improves the efficiency of the engines. Turbofan engines tend to be more efficient than turbojet engines. The design of an engine is complex and there are many considerations. In this chapter we study the overall effect of the engine assuming it was well-designed (**Figure 4.7**).

FIGURE 4.7   Turbofan engine.

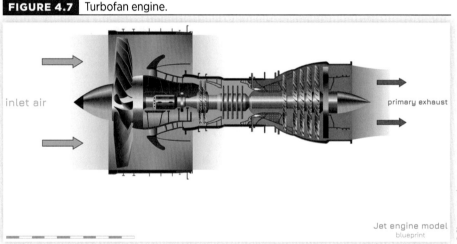

inlet air

primary exhaust

Jet engine model
blueprint

© Shutterstock

Air enters the engine with a mass flow rate of $\dot{m}_a$. Fuel is injected into the combustion canisters at a mass flow rate of $\dot{m}_f$. The exhaust gasses will have a mass flow rate of $(\dot{m}_a + \dot{m}_f)$. The velocity of the air entering the engine is the velocity of the jet aircraft, symbolized as V. The velocity of the exhaust gasses relative to the jet aircraft will be symbolized as V_e. The exhaust velocity V_e depends on the energy density of the fuel and the fuel/air mixture. The first order of business is to develop an equation for the thrust force provided by the jet engine. For steady-state flight, the thrust force of the engines must equal the drag force on the aircraft (**Figure 4.8**).

To apply the principle of conservation of momentum, we will need the absolute velocities of the air entering the jet engine and the exhaust gasses exiting the engine. The absolute velocity of the air entering the engine is zero (still air). The absolute velocity of the exhaust gasses exiting the engine is $(V - V_e)$.

FIGURE 4.8   Mass flow and velocity relative to the aircraft.

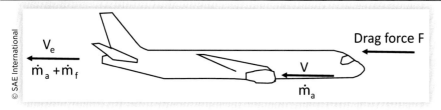

Assume that M is the mass of the aircraft and conservation of mass is being done for a small time period dt. The mass of air used during the small-time period is $\dot{m}_a dt$ and the mass of fuel used during the small-time period is $\dot{m}_f dt$. At the beginning of the time period, the aircraft (M) and the fuel ($\dot{m}_f dt$) have a velocity of V. The air ($\dot{m}_a dt$) has a velocity of zero.

$$\text{Initial Momentum} = \left(M + \dot{m}_f dt\right)V + \left(\dot{m}_a dt\right)(0) \qquad (4.21)$$

At the end of the time step, the fuel and air are consumed and have a velocity $(V - V_e)$. The aircraft still has a velocity of V. The drag force (F) acted on the aircraft a time dt. The momentum equation for the end of the time step is:

$$\text{Final Momentum} = \left(M\right)V + \left(\dot{m}_a dt + \dot{m}_f dt\right)\left(V - V_e\right) + \left(F\right)dt \qquad (4.22)$$

Setting the two equations equal, the dt divides out and the MV terms cancel. For steady-state flight the drag force must equal the thrust and it follows:

$$\text{Thrust} = \text{Drag} = \left(\dot{m}_a + \dot{m}_f\right)V_e - \left(\dot{m}_a\right)V \qquad (4.23)$$

A similar approach can be used for conservation of energy. At the beginning of the time step, the aircraft (M) and fuel ($\dot{m}_f dt$) are travelling at speed V. The air has zero velocity. Let E_f be defined as the energy density of the fuel (E_f = 46 to 47 MJ/kg for jet fuel). Let e be the thermal efficiency of the engine in converting the chemical energy in the fuel to kinetic energy of the exhaust gasses. (e is typically 45% for a modern jet engine.) The energy at the beginning of the time step is:

$$\text{Initial Energy} = \frac{1}{2}\left(M + \dot{m}_f dt\right)\left(V^2\right) + \frac{1}{2}\left(\dot{m}_a dt\right)\left(0^2\right) + eE_f\dot{m}_f dt \qquad (4.24)$$

At the end of the time step the aircraft (M) continues at speed V and the exhaust gasses exit at a speed $(V - V_e)$. The drag force F acts through a distance Vdt, so the energy consumed is F(Vdt). The energy at the end of the time step can be written as:

$$\text{Final Energy} = \frac{1}{2}\left(M\right)V^2 + \frac{1}{2}\left(\dot{m}_a dt + \dot{m}_f dt\right)\left(V - V_e\right)^2 + F\left(Vdt\right) \qquad (4.25)$$

The derivation gets a little messy at this point. Set the initial and final energy equal and expand the terms. Many terms will cancel. Then substitute for the drag force F from Eq. 4.23 and simplify. The result is a formula for the exhaust velocity V_e:

$$V_e = \sqrt{\frac{2eE_f\dot{m}_f + \dot{m}_a V^2}{\dot{m}_a + \dot{m}_f}} \qquad (4.26)$$

The thrust equation (Eq. 4.23) and the exhaust-velocity equation (Eq. 4.26) are the most important in the derivation. The air-to-fuel ratio is important for the efficiency of the engine. To have complete combustion it requires about 16 kg of air for each kg of fuel burned in the engine. The engine does not operate efficiently at the low stoichiometric ratio of 16:1. Older jet engines use about 70 kg of air for each kg of fuel burned, and the newer turbofan engines use 100 kg of air for each kg of fuel burned. The excess air reduces the exhaust temperature, and less of the energy goes out the back of the engine as heat. That is, a higher percentage of the chemical energy in the fuel is converted to kinetic energy of the exhaust gasses. Mathematically, it would be desirable to go to higher air to fuel ratios, but there are practical limitations in the combustion and distributing the heat inside the engine.

For modern jet engines the thermal efficiency is typically about 45%. There is also a thrust efficiency of the engine with a similar definition as for the propellers $2V/(V + V_e)$. The thrust efficiency is lower for a jet engine than for a propeller, but the thermal efficiency of a jet engine is higher than for the engines used to power propellers. The two factors tend to balance and yield approximately the same overall efficiency for propeller and jet-powered aircraft. The overall efficiency of a jet engine is the product of the thermal efficiency and the thrust efficiency.

$$\text{Overall Efficiency} = \frac{2eV}{V + V_e} \qquad (4.27)$$

Example 4.5: A Boeing 747 can weigh as much as 900,000 lb at takeoff. To accelerate up to speed on the runway to take off requires an initial acceleration of about 0.25 g. Assume that the engines are 45% efficient and estimate the fuel consumption during takeoff. Also, assume that the engines use 70 kg of air for each kg of fuel (**Figure 4.9**).

The mass of the 900,000 lb aircraft is 409,091 kg. Starting from rest the airplane would have zero velocity and no aerodynamic drag. There would be some rolling drag, but the main drag force is the dynamic drag required for acceleration. The required thrust for the engines is mass times acceleration.

$$\text{Required Thrust} = (409{,}091\,\text{kg})(0.25)(9.81\,\text{m/s}^2) = 1.033\,\text{MN} \qquad (4.28)$$

FIGURE 4.9 Boeing 747.

The exhaust velocity can be calculated from Eq. 4.26. The energy density of the fuel is assumed to be 46.6 MJ/kg and the thermal efficiency is assumed to be 45%. The air to fuel ratio is 70:1.

$$V_e = \sqrt{\frac{2(0.45)(46.6 \times 10^6)\dot{m}_f + 70\dot{m}_f(0^2)}{70\dot{m}_f + \dot{m}_f}} = 768.57 \text{ m/s} \qquad (4.29)$$

The required thrust value and the exhaust velocity can be substituted into Eq. 4.23 to solve for the required mass flow rate of the fuel.

$$1.033 \times 10^6 = (70\dot{m}_f + \dot{m}_f)(768.57) - (70\dot{m}_f)(0) \qquad (4.30)$$

Solving Eq. 4.30 yields $\dot{m}_f$ = 18.93 kg/s. Jet fuel has a density of 3.0 kg/gallon. The mass flow rate of fuel can be converted to a volumetric flow rate of 6.31 gal/s. A fully loaded 747 will use about 6.31 gals of fuel per second when accelerating down the runway. The required takeoff speed for the aircraft is about 200 mph, which is 89.4 m/s. At an acceleration of 0.25 g it will take 36.4 s for the aircraft to accelerate to the 89.4 m/s (200 mph) takeoff speed. The aircraft will need 1629 m, or 5344 ft of runway to get to the takeoff speed. This is a little over a mile of runway length. Boeing recommends that the runway be at least 11,000 ft, or about 2 mi, in length to ensure the aircraft has plenty of space for takeoff and landing.

Example 4.6: Assume a Boeing 747 has a weight of 700,000 lbs while cruising at 550 mph. This would be an average cruising weight for the aircraft. The aircraft has a Glide Ratio of 16. Assume that the engines are e = 45% efficient and use 70 kg of air for each kg of fuel.

   a. Calculate the drag force on the aircraft using the Glide Ratio.

   b. Calculate the exhaust velocity Ve and the fuel consumption while cruising.

   c. The fuel tank on the 747 holds 55,000 gals of fuel. Calculate the range of the aircraft.

d. Assume at the airport the cost of fuel is $4.00 per gallon. What is the cost of filling the tank?

e. If the aircraft holds 600 passengers, what is the fuel cost per passenger? What is the mpg and passenger-mpg for the aircraft? (**Figure 4.10**)

The drag force on the aircraft is estimated from the glide ratio:

$$\text{Drag Force} = \frac{(700,000\,\text{lb})(4.448\,\text{N/lb})}{16} = 194,600\,\text{N} \tag{4.31}$$

The speed of the aircraft in flight is 550 mph = 245.8 m/s. The exhaust velocity can be calculated as:

$$V_e = \sqrt{\frac{2(0.45)(46.6\times10^6)\dot{m}_f + 70\dot{m}_f(245.8^2)}{70\dot{m}_f + \dot{m}_f}} = 806.4\,\text{m/s} \tag{4.32}$$

The thrust equation (Eq. 4.23) is used to calculate the fuel consumption:

$$194,600 = (70\dot{m}_f + \dot{m}_f)(806.4) - (70\dot{m}_f)(245.8) \tag{4.33}$$

The rate of fuel consumption is $\dot{m}_f = 4.859$ kg/s, and since jet fuel has a density of 3.0 kg/gal the aircraft will consume 1.62 gal/s in steady-state flight. In comparison, the 6.31 gal/s required for takeoff you can see the aircraft uses a lot less fuel in flight. The aircraft has a 55,000-gal fuel tank, so the aircraft can fly 55,000/1.62 = 33,951 s = 9.43 h. The range of the aircraft is (550 mph) (9.43 h) = 5187 mi.

FIGURE 4.10 Boeing 747 in flight.

Los Angles to Tokyo is 5479 mi, which is further than the range calculated. Seattle to Tokyo is 4783 mi. Flying across the Pacific requires all the fuel the aircraft can hold. The density of 3 kg/gal is 6.6 lb per gal. About 55,000 gals of fuel would weigh 363,000 lb. The maximum takeoff weight for the aircraft is 900,000 lb, so the fuel represents 40% of the total weight of the aircraft on takeoff.

If the cost of fuel at the airport is $4.00 per gallon, the cost to fill the tank is (55,000 gal)($4.00/gal) = $220,000.00. With 600 passengers the cost per passenger is $366.67 per passenger. If an airline ticket to fly across the Pacific Ocean cost $2000.00, the fuel cost is a significant part of the cost of the ticket but is certainly less than half the cost.

One way to calculate the fuel economy of the aircraft is to take the range of 5187 mi and divide by the fuel consumed which is 55,000 gals. The fuel economy is 0.0943 mpg. The efficiency of large jets is typically rated in passenger-mpg, which in this case is (600) (0.0943) = 56.6 passenger-mpg. Putting two people in a car that averages 28.3 mpg on the highway yields about the same fuel economy per person as flying on a Boeing 747. Jet fuel is approximately twice the cost of gasoline. In terms of cost, one person riding in a car that averages 28.3 mpg on the highway will be about the same fuel cost per person flying on a Boeing 747.

Example 4.7: We are developing a jet airplane that will have a maximum takeoff weight of 130,000 lb and requires that the engines must be able to provide a 0.25 g initial acceleration down the runway. How much thrust is required from the engines? Assuming the engine efficiency is e = 45% and that it uses 70 kg of air for each kg of fuel burned, find the mass flow rate of fuel during takeoff (**Figure 4.11**).

The mass of the 130,000 lb aircraft is 59,091 kg. The required thrust for a 0.25 g acceleration is:

$$\text{Thrust} = (59,091)(0.25)(9.81) = 144,920 \text{ N} \tag{4.34}$$

The exhaust velocity V_e is:

$$V_e = \sqrt{\frac{2(0.45)(46.6 \times 10^6)\dot{m}_f + 70\dot{m}_f(0^2)}{70\dot{m}_f + \dot{m}_f}} = 768.57 \text{ m/s} \tag{4.35}$$

Mass flow for the fuel is calculated from the thrust formula:

$$144,920 = (70\dot{m}_f + \dot{m}_f)(768.57) - (70\dot{m}_f)(0) \tag{4.36}$$

Equation 4.36 is solved to yield a mass flow rate of $\dot{m}_f = 2.656$ kg/s for the fuel.

When cruising at 500 mph the weight of the aircraft will be reduced to an average of about 106,000 lb because some of the fuel will have been consumed. Assume a glide ration of 20 and calculate the fuel consumption while cruising.

$$\text{Drag Force} = \frac{(130,000 \text{ lb})(4.448 \text{ N/lb})}{20} = 23,574.4 \text{ N} \tag{4.37}$$

FIGURE 4.11 Jet on runway.

The cruising speed for the aircraft is 500 mph = 223.5 m/s. The exhaust velocity is calculated as:

$$V_e = \sqrt{\frac{2(0.45)\left(46.6\times10^6\right)\dot{m}_f + 70\dot{m}_f\left(223.5^2\right)}{70\dot{m}_f + \dot{m}_f}} = 799.97 \text{ m/s} \qquad (4.38)$$

Mass flow for the fuel is calculated from the thrust formula:

$$23,574.4 = \left(70\dot{m}_f + \dot{m}_f\right)(799.97) - \left(70\dot{m}_f\right)(223.5) \qquad (4.39)$$

Solving Eq. 4.39, the mass flow rate while cruising is $\dot{m}_f = 0.573$ kg/s. If the plane is to have a range of 3500 mi what should the size of the fuel tank be?

$$\text{Cruising time} = \frac{3500 \text{ mi}}{500 \text{ mph}} = 7 \text{ h} = 25,200 \text{ s} \qquad (4.40)$$

$$\text{Fuel Tank Size} = (25,200 \text{ s})\left(\frac{0.573 \text{ kg}}{\text{s}}\right)\left(\frac{\text{gal}}{3.0 \text{ kg}}\right) = 4811 \text{ gal} \qquad (4.41)$$

The fuel tank for the aircraft would need to be at least 4811 gals to give the aircraft a 3500 mi range. Fuel economy would be (3500 mi/4811 gal) = 0.729 mpg. An aircraft of this size would be able to carry about 100 passengers, so it would yield approximately 72.9 passenger-mpg.

4.4 **Fuel Economy of Rockets**

Rockets are a very small part of our transportation sector. In the grand scheme, the energy and fuel spent on rocket transportation is insignificant compared to the other methods of transportation. This section is included because many students find rockets and space exploration interesting.

Rockets must carry the fuel and oxygen to burn it which reduces the energy density of the fuel + oxygen mixture. Kerosene has an energy density of about 47 MJ/kg, but when the required oxygen is added the energy density decreases to about 10.7 MJ/kg. Liquid hydrogen has an energy density of about 130 MJ/kg, but it takes about 8 kg of oxygen to burn 1 kg of hydrogen so, after adding the oxygen the energy density drops to about 14.4 MJ/kg.

The velocity of the fuel exiting the rocket engine depends on the energy density of the fuel, and the expression can be derived using conservation of energy and conservation of momentum. Let M = the mass of the rocket and $\dot{m}_f dt$ be the small amount of fuel burned during the small amount of time dt. In this case, $\dot{m}_f$ is the mass flow rate of the fuel + oxygen mixture. Ef is the energy density of the fuel + oxygen mixture, which is 10.7 MJ/kg for the kerosene fuel and 14.4 MJ/kg for liquid hydrogen. The thermal efficiency of the rocket engine is e. The exhaust velocity V_e is the velocity relative to the rocket. The equations for thrust and exhaust velocity are below:

$$\text{Thrust} = \left(\dot{m}_f\right)\left(V_e\right) \tag{4.42}$$

$$V_e = \sqrt{2eE_f} \tag{4.43}$$

In Eq. 4.43 the E_f term is the energy density of the fuel + oxygen mixture. In addition to accelerating the rocket to speed, the rocket must overcome gravity to get into orbit. The universal gravitational constant is $G = 6.674 \times 10^{-11}$ m^3/kg s^2. The mass of the earth is $M = 5.9736 \times 10^{24}$ kg. The radius of the earth is $R = 6.371 \times 10^6$ m. Moving a mass m to a height h above the surface requires that we integrate mgh.

$$\text{Gravitational Energy} = \int_R^{h+R} \frac{GMm}{r^2} dr = \frac{GMm}{R} - \frac{GMm}{R+h} \tag{4.44}$$

If h is much larger than the radius of the earth the second term will be insignificant, and the energy required to raise a mass m = 1 kg is equal to GM/R = 62.6 MJ/kg. This would be the energy required to get beyond the gravitational pull of the earth. Since the energy density of the fuel + oxygen is 10.7 MJ/kg to 14.4 MJ/kg, it should be clear that it takes many kg of fuel to lift 1 kg of mass into deep space. The general rule of thumb is that the rocket should be 90% fuel by weight.

Example 4.8: To dock with the international space station a spacecraft must be lifted 400 km (250 mi) above the surface of the earth and must be accelerated to the orbital speed of the space station which is 27,600 kph (17,200 mph). This requires gravitational potential energy and kinetic energy. The space shuttle had a mass of 100,000 kg (220,000 lb). How much energy is required to lift the space shuttle and dock it with the international space station? (**Figure 4.12**)

FIGURE 4.12 International space station.

© Shutterstock

The gravitational energy required can be computed from Eq. 4.44:

$$\text{Gravitational Energy} = \frac{\left(6.674 \times 10^{-11}\right)\left(5.9736 \times 10^{24}\right)\left(100,000\right)}{6.371 \times 10^{6}}$$

$$- \frac{\left(6.674 \times 10^{-11}\right)\left(5.9736 \times 10^{24}\right)\left(100,000\right)}{6.371 \times 10^{6} + 400,000} = 3.696 \times 10^{11} \text{ J} \qquad (4.45)$$

The kinetic energy is ½ m V² and the orbital speed is 27,600 kph = 7666.7 m/s. Kinetic energy is calculated as:

$$\text{Kinetic Energy} = \frac{1}{2}\left(100,000\right)\left(7666.7^{2}\right) = 2.939 \times 10^{12} \text{ J} \qquad (4.46)$$

The total energy required is the sum of the gravitational and kinetic energy, which is 3.308×10^{12} J. Notice that the kinetic energy is about eight times as much as the gravitational energy. Most of the energy from the rocket engine will go towards accelerating the rocket to orbital speed for the space station. Only about 1/8 of the energy will go toward overcoming gravity. We neglected the aerodynamic drag energy required to push the rocket through the atmosphere, but aerodynamic drag energy will be smaller than the gravitational energy. The energy for aerodynamic drag in the atmosphere is insignificant compared to gravitational and kinetic energy.

If the rocket engine was 100% efficient in converting the chemical energy in the fuel to kinetic energy of the exhaust gasses, we can estimate the amount of fuel required as:

$$\frac{3.308 \times 10^{12} \text{ J}}{10.7 \text{ MJ / kg}} = 309,200 \text{ kg fuel} \quad \left(\text{for the kerosene fuel}\right) \qquad (4.47)$$

$$\frac{3.308 \times 10^{12} \, \text{J}}{14.4 \, \text{MJ} / \text{kg}} = 229{,}700 \, \text{kg fuel} \quad \left(\text{for the hydrogen fuel}\right) \qquad (4.48)$$

Please notice that the mass of the space shuttle is 100,000 kg and the mass of the fuel is two to three times as much as the mass of the space shuttle. The fuel will not be burned instantaneously as the rocket takes off. A portion of the fuel will need to be lifted and accelerated with the space shuttle. As more fuel is added to lift the weight of the fuel, it leads to the "rule of thumb" that 90% of the total weight on the launch pad needs to be fuel. When you see photographs of the space shuttle on the launch pad it is attached to a much larger rocket. Approximately 90% of the total weight on the launch pad must be fuel. That is what is required to have enough energy to lift the space shuttle into orbit and accelerate it to orbital speed so it can dock with the space station.

The rocket engine will not be 100% efficient. Around 40% efficiency is more realistic for a rocket engine. A large portion of the fuel will be needed to lift and accelerate as the rocket takes off. The fuel will be used up gradually as the rocket is in flight, and the mass of the rocket will decrease as it is in flight. We will develop a spreadsheet model of a rocket blasting off. To accelerate the rocket to the required speed it will be necessary for the mass of the rocket to be 90% fuel at takeoff. Anything less and we will not be able to get the rocket up to the required orbital speed and it will fall back to earth.

The last homework problem in the unit asks the reader to think about space travel in the practical sense. Neptune is about 2.7 billion mi from the earth. Traveling at 35,000 mph, about twice as fast as orbital speed for the space station, it would take 8.8 years to get to Neptune. The fastest space rocket we have ever made got to about 35,000 mph with the two Voyager missions. To explore our solar system in a reasonable amount of time we need to go much faster. Kinetic energy goes with the square of speed, so it takes much more energy and fuel to go faster. It is not possible to reach the speeds that we need with the current rocket fuel. We need a fuel that has a higher energy density, that is, more energy per kg. We need a lot more energy per kg. In the last homework problem, the reader is to consider the possibility of developing a nuclear rocket engine using a fuel that has a much higher energy density (hydrogen fusion). The reader will discover that even hydrogen fusion does not have enough energy to allow us to explore our solar system in a reasonable period of time.

Example 4.9: Develop a spreadsheet for a rocket going straight up. Assume that the energy density of the fuel is 10.7 MJ/kg, the efficiency is 40%, that the total weight of the rocket is 500,000 kg, and that 450,000 kg of the total weight is fuel (**Figure 4.13**).

The mass of the rocket will decrease significantly with time as the fuel is burned. Assuming a mass flow rate of $\dot{m}_f$ for the fuel, the mass of the rocket at a time t is:

$$\text{Mass} = m = 500{,}000 \, \text{kg} - \dot{m}_f t \qquad (4.48)$$

The acceleration of gravity g varies with height h above the surface of the earth according to:

$$g = \frac{GM}{\left(R+h\right)^2} = \frac{\left(6.674 \times 10^{-11}\right)\left(5.9736 \times 10^{24}\right)}{\left(6.371 \times 10^6 + h\right)^2} \qquad (4.49)$$

FIGURE 4.13   Rocket blasting off.

© Shutterstock

The exhaust velocity V_e and thrust are calculated using Eq. 4.42 and Eq. 4.43. The net lift force for the rocket is the thrust force minus the weight. Acceleration is the lift force divided by the mass.

In the spreadsheet, the approximation made is that the acceleration is constant over the time step interval chosen, which in this case is 1 s. The velocity of the rocket is incremented such that $V_{i+1} = V_i + a(\Delta t)$, where Δt is the time step. The height of the rocket is incremented according to the average velocity during the time step: $h_{i+1} = h_i + (V_{i+1} + V_i)(\Delta t)/2$.

For the spreadsheet below, it was assumed that the mass flow rate for the rocket engine was 2500 kg/s. Once the fuel is gone the rocket is done. Burning 450,000 kg of fuel at a rate of 2500 kg/s takes 180 s. At 180 s the speed of the rocket is 4933 m/s and the rocket is 232 km above the earth. This is not high enough or fast enough to dock with the space station. Working with the variables it is necessary to achieve a rocket efficiency of 80% for this rocket to be able to dock with the space station. If hydrogen fuel is used the rocket engine will need to be 60% efficient.

I hope that working through this example will illustrate to the reader the energy challenge of space flight. The fuel that we have is very marginal in providing the energy per kg mass. Because of the limitation of the fuel, nearly all the mass of the rocket must be fuel, and the rocket engines must have a very high thermal efficiency to achieve orbit. The two Voyager missions are the only two objects we have been able to push beyond our solar system, and it took decades to get them there. It was a phenomenal achievement. Flying a rocket and docking with the space station is a phenomenal achievement.

TABLE 4.3 Rocket spreadsheet.

Ef	1.44E+07	J/kg		mf	2500	kg/s			
e	0.6								
s	kg	kg/s	m/s^2	m/s	N	N	m/s^2	m/s	m
Time (s)	Mass	mass flow	gravity	Ve	Thrust	Lift	accel	Velocity	height
0	500000	2500	9.822163	4156.922	10392305	5481223	10.96245	0	0
1	497500	2500	9.822146	4156.922	10392305	5505787	11.06691	10.96245	5.481223
2	495000	2500	9.822095	4156.922	10392305	5530368	11.17246	22.02936	21.97712
3	492500	2500	9.82201	4156.922	10392305	5554965	11.27912	33.20182	49.59271
4	490000	2500	9.82189	4156.922	10392305	5579579	11.3869	44.48093	88.43409
5	487500	2500	9.821735	4156.922	10392305	5604209	11.49581	55.86783	138.6085
6	485000	2500	9.821546	4156.922	10392305	5628855	11.60589	67.36364	200.2242
7	482500	2500	9.82132	4156.922	10392305	5653518	11.71714	78.96953	273.3908
8	480000	2500	9.821058	4156.922	10392305	5678197	11.82958	90.68666	358.2189
9	477500	2500	9.820761	4156.922	10392305	5702892	11.94323	102.5162	454.8203
10	475000	2500	9.820426	4156.922	10392305	5727602	12.05811	114.4595	563.3082

Space travel is difficult. One of the major problems associated with space travel is the limited energy density of the fuel. The table above is provided to help the reader develop the spreadsheet for launching a rocket straight up into space. Setting it up like this the parameters can be changed, and the reader can explore many possibilities (**Table 4.3**).

4.5 **Homework**

1. Assume a small airplane has a total weight of 5200 lb and cruises at 120 mph. Assume that the glide ratio for the airplane is 10 at the cruising speed. The density of air is 1.17 kg/m³ at the altitude and temperature in question. The propeller has a diameter of 75 inches. Assume that the engine burns aviation fuel which has an energy density of 120,400 BTU/gal and that the thermal efficiency of the engine is 25%.

   a. Find the drag force on the airplane and the power required from the engine. (Answer: 185 hp)

   b. Find the fuel consumption in gallons per hour. If the aviation fuel has a cost of $4.35 per gallon at the airport, what is the fuel cost per hour to fly the plane? (Answer: $68.00 per hour)

   c. The fuel tank holds 65 gals. What is the range of the airplane? Please use judgment in your analysis and account for some extra fuel used in takeoff and landing (**Figure 4.14**).

FIGURE 4.14 Small airplane.

© Shutterstock

2. A small airplane must accelerate to 80 mph on the runway to takeoff. The plane has a total mass of 1200 kg. The engine and 2 m diameter propeller push the air backward at a speed V_2 = 150 mph relative to the plane when at full throttle. Assume that the aerodynamic drag force on the plane varies with the square of speed such that Aero Drag = (0.659) V_1^2, where V is in m/s and the Aero Drag force is in Newtons. Develop a spreadsheet model for acceleration of the plane down the runway at full throttle. Assume that the density of air is 1.2 kg/m³. How long does it take for the airplane to reach takeoff speed of 80 mph? (**Figure 4.15**)

FIGURE 4.15   Small airplane on the runway.

© Shutterstock

3. Assume that once the plane in problem 2 reaches 80 mph it takes off and climbs at a 10% grade. Because the plane is climbing, there is an additional gravitational force opposing the motion of 10% of the weight of the airplane. Include the gravitational force and plot the altitude of the plane with time.

4. The Learjet 60 has a maximum takeoff weight of 23,500 lb. The engines are capable of providing an acceleration on the runway of 0.35g during takeoff. Assume that the engines are 45% efficient and estimate the fuel consumption during takeoff. Besides, assume that the engines use 65 kg of air for each kg of fuel. The energy density of the fuel is 46.6 MJ/kg, and the fuel density is 3.0 kg/gallon. Calculate the gallons per second of fuel used during takeoff. (Answer: 0.232 gals per second)

5. The Learjet 60 has a weight of 22,000 lb while cruising at 484 mph. The aircraft has a glide ratio of 15 at the cruising speed. Assume the engines are e = 45% efficient and consume 65 kg of air for each kg of fuel. The density of the fuel is 3.0 kg per gallon.

   a. Calculate the drag force on the jet using the glide ratio.

   b. Calculate the exhaust velocity Ve and the fuel consumption while cruising.

   c. Calculate the fuel economy in mpg for the Learjet 60. (Answer: 2.49 mpg)

   d. The fuel tank on the jet holds 1200 gals. Calculate the range of the Learjet 60.

   e. If the maximum takeoff weight is 23,500 lb, what percentage of this weight is fuel? (**Figure 4.16**)

FIGURE 4.16   Learjet in flight.

6. We are developing a jet, that is, to have a maximum takeoff weight of 220,000 lb, and require that the initial acceleration down the runway must be 0.25 g. Energy density of the fuel is 46.6E6 J/kg and the mass density is 3 kg/gallon.

   a. Calculate the thrust required to provide the initial acceleration.

   b. Assuming the engine efficiency is e = 0.45 and that the engines use 60 kg of air for each kg of fuel, calculate the mass flow rate of fuel during takeoff. (Answer: 4.85 kg/s)

   When cruising at 450 mph the weight is 200,000 lb because some of the fuel has been burned. The engines use 90 kg of air for each kg of fuel while cruising. Assume a glide ratio of 20 for the aircraft and calculate the fuel economy while cruising at 450 mph. If the plane is to have a range of 3000 mi, how large should the fuel tank be, and what is the weight of the fuel? (Answer: 0.389 mpg, 7704 gals, 50,800 lb, or 23.1% of the vehicle weight is fuel)

7. Assume that we plan to fire a rocket to lift a spacecraft into space using a rocket fuel/oxygen mix that has an energy density of 14.4 MJ/kg. The spacecraft is approximately the size of the space shuttle, which had a mass of approximately 100,000 kg. Assume 45% thermal efficiency for the rocket engines.

   a. The "rule of thumb" is that a rocket + spacecraft should be 90% fuel by weight. Assume that the rocket itself is 95% fuel by weight and that it must lift the 100,000 kg spacecraft. How much fuel will be in the rocket, and what is the total weight of the rocket and spacecraft?

   b. Develop the spreadsheet like the one developed in the book for this rocket going straight up (**Figure 4.17**).

FIGURE 4.17   Space Shuttle launch.

© Shutterstock

8. Space travel with conventional rocket fuel is difficult. The energy density of the fuel is much too low to go very far into space. We made it to the moon and back, and we might make it one way to Mars, but beyond that we will need a fuel with a higher energy density.

For this problem assume that we develop the technology to control a nuclear fusion reaction converting Deuterium-Deuterium hydrogen into helium. The energy density of the fuel is 250×10^{12} J/kg. Assume that we have a spacecraft that has a total mass of 500,000 kg and is 50% fuel by mass. Thermal efficiency of the rocket engines is 45%. Blasting this rocket off would be a larger energy release than all the hydrogen bombs ever created, and it might destroy life on a large part of the earth. But overlooking that problem, use the spreadsheet approach to calculate the speed of the rocket when it has burned half of its fuel supply. How does that speed compare with the speed of light? (This is a thought problem to help you understand that even with the most powerful fuel we know of, nuclear fusion, there is not enough energy to accelerate the spacecraft to speeds necessary for space exploration. We would need to discover a fuel with a much higher energy density than nuclear fuel.)

Reference

1.   https://en.wikipedia.org/wiki/Lift-to-drag_ratio#Glide_ratio

Unit 2
Energy and the Environment

As a society we have come a long way since colonial times. The Industrial Revolution started in the early 1800s, and all the inventions and technology dramatically improved the quality of our lives. We have hot and cold running water. Our homes and buildings are climate controlled. We have electric power available for the multitude of electronic devices we use. We have a safe food supply and can buy almost any food we want at any time of the year. We have a transportation system that allows us to travel long distances with ease and in comfort. We have modern medicine that can often heal us when we have a medical problem. Our lives are better and easier than at any time in history.

The modern lifestyle we lead has an impact on the environment. Maintaining our lifestyle requires that we use a very large amount of energy. When quantifying the amount of energy used the numbers are so large that it is difficult to understand what the numbers mean. The first part of Unit 2 will help the reader comprehend and make sense of the amount of energy used in the United States and in the world. Burning fossil fuels adds carbon dioxide to the atmosphere, and scientists believe that the carbon dioxide in the atmosphere contributes to global warming. In Unit 2, the book will show the reader how to calculate carbon footprint and carbon emissions from fossil fuels. The book also illustrates how to estimate other pollutants that come from burning fossil fuels including sulfur dioxide, nitrous oxides, carbon monoxide, and particulate emissions. The reader will learn to do the calculations and will be able to show quantitatively that burning coal generates more pollution and carbon emissions than burning natural gas. Burning propane, diesel fuel, gasoline, or jet fuel is cleaner than burning coal, but not as clean as natural gas. The focus in Unit 2 is on how to do the calculations associated with burning fossil fuels. A relatively simple radiative heat transfer model is developed for the earth so the reader can calculate how increasing the atmospheric content of carbon dioxide will increase the average temperature of the earth.

In order to reduce our carbon footprint, it will be necessary to produce the electricity we use from renewable sources: hydro, wind, and solar. Unit 3 shows how to estimate the amount of electric energy that can be produced from these sources so the reader can estimate how many wind turbines, solar panels, dams and lakes are necessary to produce the amount of electricity we will need. Generating the energy we need from renewables is not going to be easy and not going to happen in the near future. Batteries are an essential part of making the renewable energy system work, and the batteries we have today are not adequate. Vehicles will require lightweight batteries that can store a lot of energy. Utilities will require huge batteries to allow them to keep the electric grid balanced and handle peak loads. The last part of Unit 2 discusses the batteries that are currently available and their strengths and weaknesses.

United States and World Energy Consumption

Human innovation, machines, and other technology allow us to live the modern lifestyle we have today. We tend to think about the technology because that is what we see and what impacts our day to day life. But in the background, it takes a lot of energy to power all of the inventions and technology. Before the industrial revolution most of our energy came from human and animal power. Horses and dogs were important, because they took some of the workload of humans and made our lives better than what we had before. The steam engine and electric power plants gave us more energy to work with and made our lives much better than what we had with human and animal power. The Lawrence Livermore National Laboratory keeps track of the energy used in the United States and the results are summarized in the chart below [1]. A Quad is defined as a quadrillion BTU or 1×10^{15} BTU. In the USA we have used about 100 Quads annually for each of the last 10 years.

In **Figure 5.1** the left column is the primary sources of energy we use. The first primary source of energy is solar energy, and it is important to note that the Department of Energy only considers the solar voltaic panels and the solar hot-water heaters in this category. Solar energy is mainly used in agriculture, as it is used to grow the food we eat. If agriculture is included, solar is our largest and most important energy source, and we use more solar energy than all the others combined.

We often think of electricity as an energy source, but there is no naturally occurring electricity we can harness. Electricity must be generated from the primary energy sources. Of the 101.2 Quads of energy used in 2018, 38.2 Quads were used to generate electricity. Generating electricity is the largest single use of energy.

Electricity is important because it is the most efficient way to transmit and distribute energy. We can transmit electricity over long distances and distribute it to the customers who need

FIGURE 5.1   Estimated US Energy Consumption in 2019.

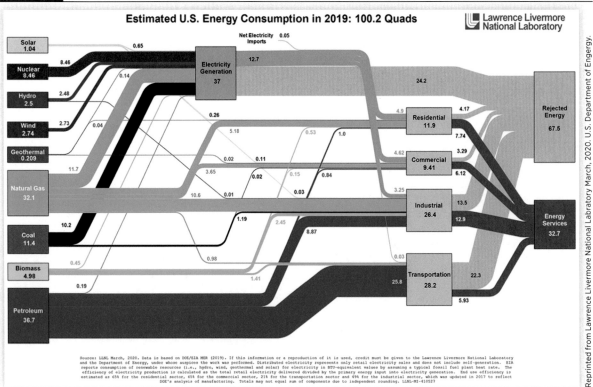

electricity with very little energy loss. It is the energy efficiency associated with the transmission and distribution of the energy that makes electricity so valuable to us.

The early powerhouses had shafts and pulleys inside a building, and equipment was powered using a belt drive system from one of the overhead pulleys. It was dangerous and had low-energy efficiency. One could envision a mechanical energy distribution system for a city that had shafts, gears, and universal joints on top of telephone poles through the city. The shaft system would go to houses and power overhead shafts and pulleys that could drive equipment in the houses using a belt system like in the old powerhouses. A mechanical distribution system like this would be very expensive and inefficient. There would be large energy losses in all gears, universal joints, pulleys, and belts. It is a ridiculous idea but visualizing a mechanical distribution system helps you appreciate the value of the electrical distribution system. It is not that electric energy has a high value, but that we can transmit and distribute the electricity very efficiently and at a low cost.

Looking at Figure 5.1 we can draw some conclusions about the energy used to generate electricity. All the primary sources of energy are used in generating electricity. A question that always comes up is "What is biomass?" Biomass includes the alcohol mixed with gasoline, the biodiesel, and the burning of wood or other organic materials. It is fuel made from plants. The table below summarizes how the different types of primary energy are used (**Table 5.1**).

If we divide the electricity production by types, fossil fuel contributes 23.34 Quads or 61.0% of the electricity produced. Renewables contribute 6.46 Quads or 16.9%, and

TABLE 5.1 Primary energy used to generate electricity in USA.

Primary Energy	Amount (Quads)	Percentage (%)	Type
Solar	0.61	1.6	Renewable
Nuclear	8.44	22.1	Nuclear
Hydro	2.67	7.0	Renewable
Wind	2.53	6.6	Renewable
Geothermal	0.15	0.4	Renewable
Natural Gas	11.0	28.8	Fossil Fuel
Coal	12.1	31.6	Fossil Fuel
Biomass	0.5	1.3	Renewable
Petroleum	0.24	0.6	Fossil Fuel
Totals	38.24	100.0	

Nuclear contributes 8.44 Quads or 22.1%. Some people may want to put the nuclear energy in with renewables since we have an almost unlimited amount of nuclear fuel. Compared with the 2016 data, renewables have increased by about 2% and fossil fuels have decreased by about 2% in the mix of generating electricity.

As most of the electric energy comes from heat engines, the second law of thermodynamics says that most of the energy will become heat that goes into the atmosphere, which is called "Rejected energy" in Figure 5.1. The "Rejected energy" and "Energy services" division on the right side of the chart can be used to assess our overall energy efficiency as a nation. Of the 101.2 Quads of energy used, 68.5 Quads (67.8%) are expelled to the atmosphere as wasted heat and 32.7 Quads (32.3%) become useful energy for us. Our overall efficiency is 32.3% in using our primary energy.

Figure 5.1 divides the energy we use into categories of residential, commercial, industrial, and transportation. Residential, commercial, and industrial customers use a lot of energy heating our homes and buildings in the winter, commercial cooking and industrial canning, and the melting and processing of plastics, metals, and ceramics. When we need heat, we can use fossil fuel, especially natural gas, very efficiently to provide the heat. Modern furnace systems are more than 90% efficient in using the heat energy in the natural gas to heat our homes. Using the natural gas directly is much more efficient overall than using electricity for heat. If we use natural gas in a power plant to make electricity, only about one-third of the energy in the natural gas will be converted to electricity. The remaining two-thirds will be lost as heat in the atmosphere. An electric furnace may be 100% efficient in converting the electricity into heat but, when you look broadly, we only recover about 33% of the heat energy originally in the natural gas. It is much more efficient to use the natural gas directly in the furnace. This is true for almost any application where we need heat; it is most efficient to use the natural gas (or other fossil fuel) directly to generate the heat. There are cases where we only need a small amount of heat and using an electric heater makes sense, but in general using electricity to make heat is inefficient and wasteful. Electricity should be used to power electric motors and other electrical and electronic devices where we need electricity.

Figure 5.1 also provides information about the efficiency in generating electricity and in providing the energy for residential, commercial, industrial, and transportation applications. **Table 5.2** shows a summary.

TABLE 5.2 Summary of energy use and efficiency in electricity, residential, commercial, industrial, and transportation.

	Useful energy (Quads)	Total energy (Quads)	Efficiency (%)
Electricity	12.9	38.2	33.8
Residential	7.72	11.9	64.9
Commercial	6.14	9.45	65.0
Industrial	12.9	26.3	49.0
Transportation	5.95	28.3	21.0

© SAE International

The efficiencies for residential, commercial, and industrial customers are much higher than for generating electricity and transportation. In residential, commercial, and industrial applications a significant percentage of the energy in the fuel is used to make heat, which can be done at a high efficiency. For generating electricity and transportation, the energy in the fuel is used in a heat engine (IC engine or steam turbine), which has a much lower efficiency than using the energy in the fuel directly as heat. The overall efficiency in generating transportation is only 21%, which means that 79% of the energy in the fuel (mostly gasoline, diesel, and jet fuel) is dumped into the atmosphere as heat.

The next step is to gain a deeper understanding of how much energy is used in the USA. We use 101.2 Quads of energy annually, but most of the people have a difficult time in understanding the magnitude of that amount of energy. There are 327.2 million people in the USA, so each person would use 309.3 million BTUs annually. Dividing by 365 days in a year yields that each person uses 847,342 BTUs daily. This is the average daily energy consumption for people in the USA.

Before the industrial revolution, horses were used to provide some of the energy that we use. We might consider a horse that could produce one horsepower for 8 h per day. This would be a very hard-working horse. Horses have personalities and convincing the horse to work this hard for 8 h per day would be a challenge. But, in an ideal situation, a horse could produce 8 horsepower-hours of energy in a day. We need to convert the energy into BTUs for comparison.

$$(8 \text{ hp-h})\left(\frac{550 \text{ ft-lb} / \text{s}}{\text{hp}}\right)\left(\frac{\text{BTU}}{778.2 \text{ ft-lb}}\right) = 20,355 \text{ BTU} \tag{5.1}$$

Dividing the 847,342 BTU the average person uses each day by the 20,355 BTU a horse could provide in a day yields 41.6, which means we would need 41.6 horses per person in the USA to provide the energy that we use. We would need to figure out a way that each person could work 41.6 horses for 8 hours a day. For our nation we would need about 13 billion horses to provide the energy that we need. It would not be possible for us to grow enough food to feed them, and I am not sure what we would do with the manure.

Another way to look at the energy we use is to consider how we could replace it with human power. A well-conditioned human can provide about 10% of a horsepower. It would be hard for any person to do that for 8 hours a day, but even in the most idealistic case it would take 400 humans to provide the energy necessary for one person to live the modern lifestyle that we have. If all our energy came from human power, we would have to live a much simpler life.

The industrial revolution brought us many inventions and technologies that improved the quality of our lives. As the inventions came, we gradually increased our energy consumption every year. Today, we use a tremendous amount of energy, an amount that would be unimaginable to people who lived before the industrial revolution. To maintain our modern lifestyles, we will need to continue using these large amounts of energy and improve the energy efficiency of the devices we use. Using large amounts of energy is creating environmental problems and there is no easy answer to addressing these problems. None of us want to give up our modern lifestyle, which means we cannot reduce the amount of energy we are using. Most of our energy comes from fossil fuels and it will have to continue that way for the near future. We are working on developing energy sources that are more environmentally friendly.

Burning fossil fuels generate carbon dioxide which is released into the atmosphere, and the carbon dioxide is causing a greenhouse effect that causes our planet to warm. The environmental protection agency (EPA) tracks greenhouse gasses for the USA [2]. **Figure 5.2** is from their 2019 report, which shows results through 2017.

A metric ton is 1000 kg, and the USA generated 5270.7 million metric tons of carbon dioxide in 2017. In 2017, 4912.0 million metric tons of the carbon emissions came from fossil fuels (natural gas, coal, and petroleum).

Of the carbon dioxide emissions from fossil fuel in 2017, 35.8% came from coal, 29.5% from natural gas, and 44.7% from petroleum. Agriculture and industrial processes produce small but significant amounts of carbon dioxide. Generating electricity produces 35.3% of the carbon emissions, transportation 36.7% and smaller amounts come from residential, commercial, and industrial uses. In reducing our carbon emissions, most of the effort is focused in generating electricity and transportation because these two sources generate 72% of the carbon emissions.

FIGURE 5.2   Gross Greenhouse Gas Emissions USA.

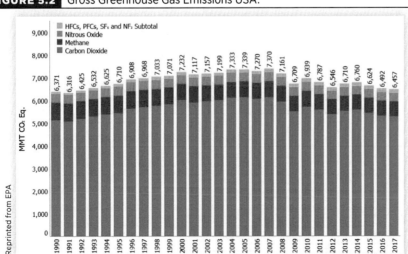

FIGURE 5.3 Estimate USA Water Usage 2005.

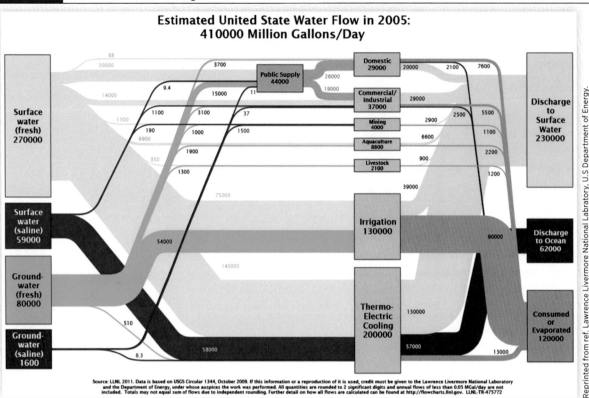

The Lawrence Livermore National Laboratory also tracks our water usage [3]. Water is not energy but is important, and the national lab produces a nice chart illustrating our usage of water (**Figure 5.3**).

Our largest source of water is surface water, which is drawn from lakes and rivers. Surface water accounts for about two-thirds of the water we use. The second largest amount is ground water, which is the water drawn from wells. Much of the population lives near the ocean and can use surface water (saline), which is ocean water for some uses. There area small number of saline wells.

Largest use of water is for thermo-electric cooling which, for the most part, is in electric power plants. Power plants use a lot of water. Cooling towers used to provide chilled water for cooling buildings also use water and fall into this category. Most of the fresh water used flows back into the rivers and lakes or, in the case of saline water, flows back into the ocean. There is significant evaporative loss in the cooling towers. Irrigation is the largest consumer of water in the sense that most of the water used for irrigation flows into the soil or evaporates. About 70% of the water used for irrigation is consumed. A significant fraction of the water consumed in irrigation flows through the soil and recharges the groundwater. It is difficult to quantify exactly how much, but some of the water classified as consumed in irrigation will become groundwater.

The next step is to look at the energy used in the world. **Figure 5.4** illustrates world energy consumption since 1990. The world used about 550 Quads of energy in 2018. There was a time when the United States used about one-fourth of the energy in the

FIGURE 5.4   World and regional energy consumption.

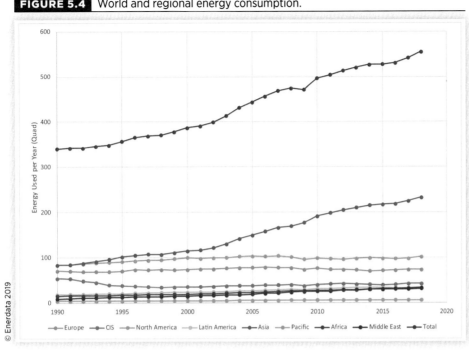

© Enerdata 2019

world, but that has gradually been reduced to about 18% as the rest of the world has modernized. The United States represents about 5% of the world population using 18% of the energy, so on a per-person basis we use a lot of energy. Data for **Figures 5.4, 5.5, and 5.6** were obtained from Enerdata's Global Energy Statistical Yearbook [4] at https://www.enerdata.net/publications/world-energy-statistics-supply-and-demand.html.

FIGURE 5.5   Energy consumption in CIS, Latin America, Africa, Middle East, and Pacific.

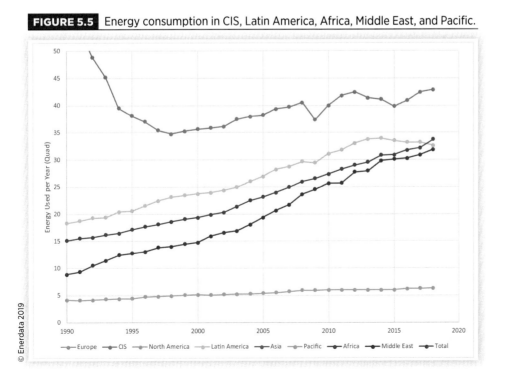

© Enerdata 2019

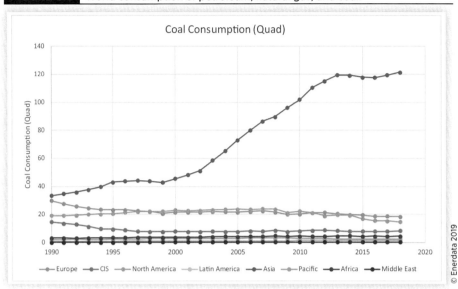

FIGURE 5.6 World consumption of petroleum, natural gas, and coal.

The data from Enerdata gives petroleum (oil) consumption in millions of metric tons, natural gas in billions of cubic meters, and coal in millions of metric tons. To develop Figure 5.4 I had to convert these numbers to Quads of energy, which requires conversion factors. I searched the literature and used average values that I found for the conversion factors. For petroleum I assumed a million metric tons was equal to 0.0458 Quad of energy. For natural gas I assumed a billion cubic meters was equal to 0.0353 Quad of energy. For coal I assumed a million metric tons was equal to 0.0222 Quad of energy. Figure 5.4 is based on these assumptions.

Energy usage in North America and Europe has been consistent over the past 30 years. There have been significant improvements in efficiency in electric power production, transportation, furnaces, and most of the other devices that we use. Our increase in energy needs have been offset by improvements in energy efficiency, so that overall energy usage has remained consistent. Energy usage is likely to gradually increase as the populations grow. The year 2009 was a bad year economically for most of the world, and most parts of the world showed a downward tick in energy consumption in 2009.

The large increase in energy usage has been in Asia, as shown in Figure 5.4. The data for Asia does not include commonwealth of independent states (CIS) (Russia and some of the former Soviet bloc countries), even though they are technically in Asia. The data for Asia includes China, India, Japan, and other countries in the Far East. About half of the world population lives in Asia, and they made great progress in modernizing and improving their lifestyles. Increase in energy consumption in this part of the world will continue to increase for at least a few more decades as they raise their quality of life to be comparable to the US and Europe.

The lines for Latin American, Middle East, and Africa on the chart appear to be flat, but this is because of scaling of the chart. In Figure 5.5 the *y*-axis was adjusted to show better how energy usage is changing in other parts of the world.

Energy use in the Middle East and Africa has grown steadily in the last 30 years. Energy use in Latin America grew steadily until about 5 years ago. Political issues and violence in Latin American countries caused the economies in some countries to decline, leading to less energy consumption. The CIS line in the chart is Russia and some of the former Soviet Union countries. There were political issues in the 1980s and 1990s which led to the breakup of the Soviet Union. The economy in that part of the world collapsed. We did not hear much about it at the time, but the people in that part of the world suffered and the economy has never really recovered. The Pacific line on the chart includes Australia, New Zealand, Indonesia, and the island countries in that part of the world. Energy consumption has increased steadily in that part of the world.

Another way to look at world energy consumption is by the type of energy used. Figure 5.6 illustrates world energy consumption by the type of fossil fuel used. Petroleum was our largest source of energy until 2005 when coal became the largest source of energy in the world. Nuclear and renewable energy are much smaller sources in comparison and are not plotted in Figure 5.6. Renewable energy sources have increased significantly every year. Nuclear energy has declined as old nuclear power plants have been closed.

References

1. https://flowcharts.llnl.gov/
2. https://www.epa.gov/sites/production/files/2019-04/documents/us-ghg-inventory-2019-main-text.pdf
3. https://flowcharts.llnl.gov/commodities/water
4. https://www.enerdata.net/publications/world-energy-statistics-supply-and-demand.html

Combustion of Fossil Fuels

6.1 Introduction

When we burn fossil fuels the combustion products are primarily water vapor and carbon dioxide, which go into the atmosphere. Carbon dioxide is a greenhouse gas, and it is linked to global warming. Fossil fuels also produce significant amounts of sulfur and trace impurities of mercury, lead, and other heavy metals. This is especially true for coal. There are things that can be done at the power plant to remove heavy metals like lead and mercury, but the sulfur is difficult to remove. Most of the sulfur will go into the atmosphere as sulfur dioxide, which is the primary source of acid rain.

Electric vehicles are pollution free at first glance, but most of the electricity is produced using fossil fuels. To be fair in evaluating electric vehicles, we should include the pollution and carbon emissions associated with producing the electricity the vehicles use. The approach used in this book will be to look at the percentage of electricity generated by fossil fuels and get an average value of carbon emissions and pollution generated per kW-h of electric energy. From there we can estimate the amount of carbon dioxide and other pollutants that are generated indirectly from an electric vehicle. We will find that electric vehicles are low emission vehicles but are not zero emission vehicles. Small hybrid-electric cars have essentially the same emissions per mile traveled as small electric cars.

Wind turbines, solar panels, and hydroelectric dams are considered pollution free methods of generating electricity. But the production of wind turbines, solar panels, and hydroelectric dams requires that a lot of fossil fuel be used for mining, manufacturing, transportation, and maintenance. It is difficult to account for the fossil fuel used in production and we will not try to account for it in this book.

In this unit we will study the chemistry of combustion of fossil fuels. The easiest fuel to deal with is natural gas, so we begin with natural gas. Natural gas is composed mostly of methane CH_4. We visualize methane in the figure below. The carbon atom has four covalent bonds and each of the hydrogen atoms have one covalent bond. On paper we draw it in two-dimensions but, in reality, it is a three-dimensional molecule. The four hydrogen atoms are arranged as the corners of a tetrahedron or pyramid around the carbon atom as illustrated in **Figure 6.1** on the right rather than in the plane as we draw it on paper as illustrated on the left.

There are measurable amounts of ethane and propane in the natural gas. Natural gas will be at least 95% methane, but it is not 100% methane. We can envision the ethane and propane molecules as shown in the figures below. In reality, they are three-dimensional molecules that are twisted and not flat the way we draw them on paper, but the two-dimensional drawings are helpful (**Figure 6.2**).

In addition to ethane and propane there are smaller amounts of CO_2, N_2, H_2S, He, particulate material, and trace amounts of other materials. We clean and purify the natural gas as best as we can, but it will never be 100% pure. Since natural gas is primarily methane, the combustion associated with methane is a good approximation for combustion of natural gas. The chemistry equation for complete combustion of methane is shown in Eq. 6.1 below:

$$CH_4 + 2O_2 \rightarrow CO_2 + 2H_2O \qquad (6.1)$$

The equation assumes that natural gas is all methane and that we achieve complete combustion. Air is 78.09% nitrogen, 20.95% oxygen, 0.93% argon, and trace amounts of other gasses. The percentages are by volume; the mass percentages are different. In an ideal gas each molecule takes up the same amount of volume regardless of its weight. For each 100 molecules in air we could approximate the chemical composition of air with the following formula:

$$Air = 78N_2 + 21O_2 + Ar \qquad (6.2)$$

FIGURE 6.1 Molecular structure of methane.

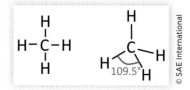

© SAE International

FIGURE 6.2 Ethane and Propane molecules.

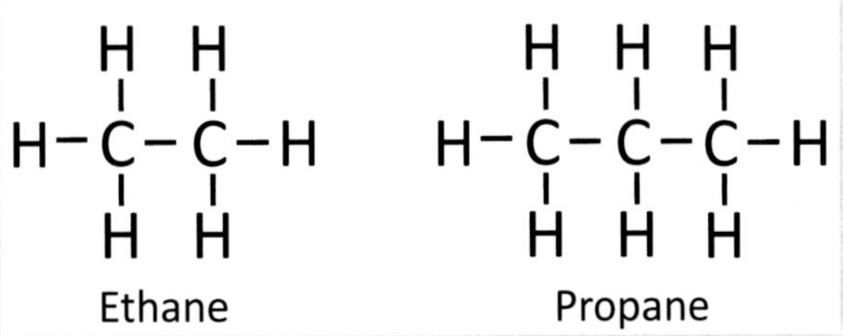

© SAE International

Chemical reactions are based on probability and there are many molecules involved in the chemical reaction. There will be trace amounts of many different chemicals, but there is a strong driving force toward producing carbon dioxide and water. In a good combustion process, we will achieve near complete combustion, but we can never achieve 100% complete combustion. There will be small amounts of carbon monoxide, and since the air used for combustion contains nitrogen there will be small amounts of nitrous oxide created during combustion. The hydrogen sulfide impurity in the natural gas will create some sulfur dioxide, and the particulate matter will not burn. There are trace amounts of other chemicals, but most of the combustion products are carbon dioxide and water.

For a good combustion process, we can estimate the pollution generated from burning natural gas. For each billion BTU of energy we need from the gas, we will generate approximately [1]:

- 40 lb carbon monoxide (CO)
- 1 lb Sulfur dioxide (SO_2)
- 92 lb Nitrous Oxide (NO_x)
- 7 lb particulate

We don't want any of these things in the air we breathe, but it is a part of the process of using natural gas. Carbon monoxide is poisonous. Sulfur dioxide causes acid rain and is mildly poisonous. Nitrous oxides cause smog and are not good for our health either. Breathing the fine particles in the particulate matter causes lung problems. Some of the particulate matter is carcinogenic. Natural gas is cleaner than other fuels that we burn, but it does generate some pollution.

Most of the chemical reaction of burning natural gas will be complete combustion of methane. Water and carbon dioxide are produced from combustion. Both water and carbon dioxide are greenhouse gasses in the sense that they absorb radiation, but the water doesn't stay in the atmosphere long. It falls as rain. Carbon dioxide stays in the atmosphere longer. Plants gradually absorb carbon dioxide from the atmosphere using photosynthesis.

6.2 **A Theory for Global Warming**

Global warming is a topic that is in the news and is controversial. I'm not trying to advocate for one side or the other in this book. My goal is to teach the science involved and how to do the calculations. You should make your own decision as to whether or not you believe global warming is caused by human activity.

The earth has gone through warmer and cooler periods in its history. There have been times when the earth was much warmer and when it was much cooler than it is today. Geologists study the history of the earth, and the theory that scientists are using to explain global warming comes out of geology. It is sometimes referred to as the "snowball earth" theory [2]. It is a good theory, but it is just a theory. It is not necessarily correct, and not all scientists agree with the theory. Other theories involve the change in earth's rotational orbit known as Milankovitch cycles [3], changes in volcanic activity on the planet, changes in the intensity of the sun, and movement of the tectonic plates on the earth.

Most of the land area is in the northern hemisphere: North America, Europe and Asia. In warmer periods these land masses can support a lot of plant life, and the plants remove carbon dioxide from the atmosphere through photosynthesis. In colder periods the land masses are covered with snow and ice and there is very little plant life to remove carbon dioxide from the atmosphere. The oceans will absorb and release carbon dioxide. Equilibrium concentration in the ocean depends on the partial pressure of carbon dioxide in the atmosphere.

To illustrate the cycle, we will start with a cold period on earth, where a large fraction of the land masses is covered in ice like an ice age. The ice reflects sunlight, reducing the solar energy absorbed by the earth. We will start with the carbon dioxide concentration in the atmosphere and oceans at a low level.

During the cold period the volcanic activity puts carbon dioxide into the atmosphere faster than the plant life can remove the carbon dioxide. Volcanic activity is not just the eruptions of volcanos. Yellowstone National Park adds a tremendous amount of carbon dioxide to the atmosphere each year from its hot springs and other thermal features. The volcanic activity in Hawaii is always adding carbon dioxide to the atmosphere. There are many places on the planet that have active volcanoes or other volcanic activity adding carbon dioxide to the atmosphere.

As the carbon dioxide builds in the atmosphere it absorbs some of the infrared radiation that the earth radiates into space. The "greenhouse" effect of the carbon dioxide in the atmosphere helps the earth begin to warm. As the partial pressure of carbon dioxide increases in the atmosphere, the oceans will absorb some of the carbon dioxide because it is soluble in water.

Melting the ice is a slow process because of the "heat of melting" phase change energy of the ice. It takes approximately 145 BTU to melt one pound of ice. The definition of a BTU is the energy to raise one pound of water 1°F. Converting the ice to water takes an additional 145 BTU per pound over and above the energy required to warm it to the melting temperature. It is the phase change energy of ice that makes it so great for cooling our sodas and for keeping things cool when we go camping. The earth will start to warm and melt the ice, but it will take many thousands of years to melt the ice from the large land masses in North America, Europe and Asia. The carbon dioxide concentration in the atmosphere will continue to rise during this time period and will reach a relatively high value. The ocean will continue to absorb carbon dioxide too.

As the ice recedes plant life will increase, and the plants will remove carbon dioxide from the atmosphere through the photosynthesis process. At some point the plants will remove more carbon dioxide from the atmosphere than is added by the volcanic activity, and carbon dioxide levels will begin to decline in the atmosphere and in the ocean.

As the carbon dioxide levels decline, the earth will begin to cool, but forming the ice at the poles is a slow process that will take many thousands of years. During that period the plants will continue to thrive and remove carbon dioxide from the atmosphere and the carbon dioxide concentration in the atmosphere and ocean will decline to a relatively low value.

As the ice advances from the poles toward the tropics the plant life is killed, and the plants are buried. Over a long period of time the buried plants become the fossil fuel that we are using today. We talk about carbon sequestration today and envision methods of storing the carbon dioxide somehow rather than releasing it into the atmosphere. The natural process has been for the plants to take the carbon dioxide out of the atmosphere and then be buried in the ground, effectively storing the carbon in the ground. As we burn fossil fuel, we are releasing the carbon dioxide back into the atmosphere. Most scientists are concerned that this will lead to very high levels of carbon dioxide in the atmosphere and oceans, which will lead to an exceptionally warm period on earth.

An exceptionally warm period on earth may lead to many environmental consequences. It will lead to melting of the ice in the polar regions, which will cause the oceans to rise and damage our coasts. The coastal areas are very valuable property and we will not give them up easily. There will be many public works projects directed at protecting our coasts, which will create lots of jobs in the construction industry. The poorer and low-lying countries will not be able to afford to protect their coasts and will lose a lot of property.

A warmer atmosphere will have more moisture and energy. Storms will be stronger and more frequent. The additional rain will help with drought in some parts of the world and will cause more flooding in other parts. Some parts of the world will benefit from a warmer atmosphere and some will be hurt.

The warming of the planet will not be evenly distributed. Models project a large amount of warming in the polar regions and a smaller amount of warming in the tropics. Some areas will benefit from global warming because it lengthens their growing season

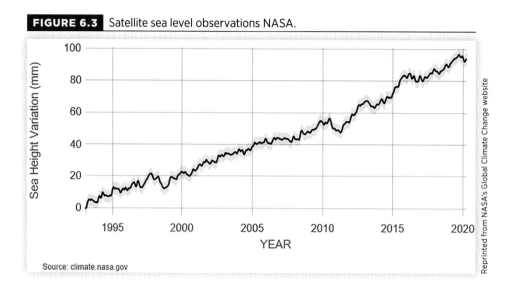

FIGURE 6.3 Satellite sea level observations NASA.

Source: climate.nasa.gov

and allows them to grow more food. Winters will be less harsh in the high latitudes and polar regions.

The carbon dioxide content of the ocean will rise, and we see that this is already killing a lot of the coral in the ocean. It is hard to assess the full impact of the increase of carbon dioxide in the ocean.

We have enough data to show clearly that the oceans are rising. **Figure 6.3** is a plot of satellite data taken by NASA and is available at Ocean Level Data [4] https://climate.nasa.gov/vital-signs/sea-level/. The graph is based on satellite data starting in 1993. It shows the ocean has been rising on average 3.3 mm annually.

Figure 6.4 is based on tidal data [4], which has been recorded since 1880 and is available at the same web site as Figure 6.3. Measuring ocean level to a tenth of a millimeter is difficult and there is obviously some noise in the data, but the trend is clearly upward. Since 1993 the ocean has risen about 90 mm, which is about 3.5 inches. A rise of 3.5 in. in 25 years is substantial. There is a lot of noise in the tidal data, but since 1880 the oceans have risen about 230 mm, which is about 9 in.

The purpose of presenting the sea level data is to show that we have data clearly establishing that the oceans are rising. Part of the rise in ocean depth is due to thermal expansion of the water in the oceans due to temperature increase in the oceans. Part of the rise in ocean depth is due to melting of ice in the polar regions on earth.

Figure 6.5 shows the rise in carbon dioxide in the atmosphere since 2005 [5]. On average carbon dioxide has been increasing in the atmosphere at about 2.2 ppm each year since 2005. The data for this chart was collected by NOAA.

Figure 6.6 shows a much longer historical record of carbon dioxide in the atmosphere obtained from ice core samples [5]. As the earth has gone back and forth between

FIGURE 6.4   Seal level ground data.

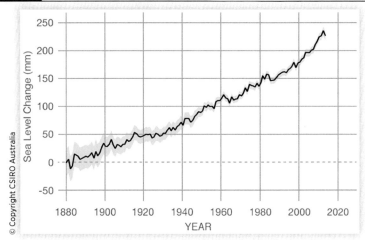

FIGURE 6.5   Atmospheric carbon dioxide since 2005.

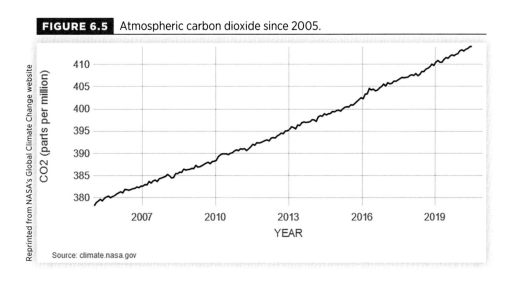

Source: climate.nasa.gov

warmer and cooler periods the carbon dioxide has gone back and forth between lower and higher values. A typical cycle lasts about 100,000 years, as shown on the chart. The warmer periods on earth correspond to times where the carbon dioxide was at relatively high values, and the cooler periods correspond to times when the carbon dioxide was at lower values. Since 1950 the earth has been at historic high values, much higher than any time in the last 400,000 years. Most scientists believe this historic high carbon dioxide reading in the atmosphere will lead to a very warm period in the history of the earth.

FIGURE 6.6 Carbon dioxide measurements from ice core samples.

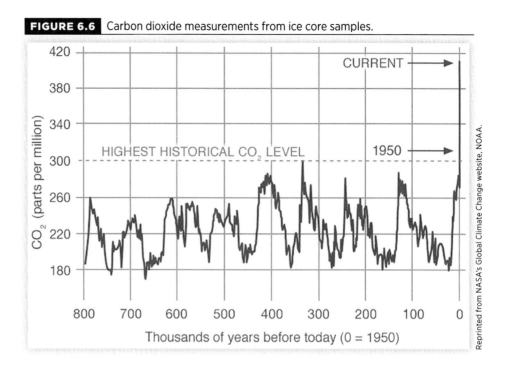

Figure 6.7 shows an average global land-ocean temperature index [5]. There is a lot of noise in this data because measuring the average earth temperature is a difficult thing to do, but the trend is upward. The average global temperature is up approximately one degree Celsius over the last 100 years.

FIGURE 6.7 Global land-ocean temperature index.

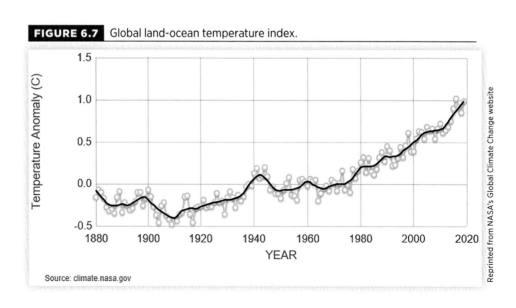

Source: climate.nasa.gov

6.3 **Combustion of Natural Gas**

In this chapter we will study the combustion of fossil fuels with the focus on estimating the pollution and carbon emissions for different types of fossil fuels. The first step is to analyze the combustion of natural gas in more detail as a prelude to analyzing more complicated fossil fuels. The basic equation for combustion of methane is:

$$CH_4 + 2O_2 \rightarrow CO_2 + 2H_2O \tag{6.3}$$

The molecular weight of carbon is 12, meaning one mole (Avogadro's number) of carbon atoms will have a mass of 12 g. Hydrogen has a molecular weight of 1 and oxygen has a molecular weight of 16. In this book I am rounding the molecular weights to whole numbers. In reality the average carbon atom has a molecular weight slightly higher than 12 but using the exact numbers doesn't make much difference in the final results. The molar weights of the compounds can be calculated as:

$$\text{Molar weight } CH_4 = 12 + 4(1) = 16 \, g / mole \tag{6.4}$$

$$\text{Molar weight } CO_2 = 12 + 2(16) = 44 \, g / mole \tag{6.5}$$

$$\text{Molar weight } H_2O = 2(1) + 16 = 18 \, g / mole \tag{6.6}$$

There is an important conclusion to this analysis. When we examine the molar weights and the chemical reaction, we can conclude that burning 16 g of methane (CH_4) will produce 44 g of carbon dioxide (CO_2) and 36 g of water (H_2O). The reader should look at Eqs. 6.3 through 6.6 and make sure they understand. These are important conclusions.

Natural gas typically has 950–1050 BTU of energy per standard cubic foot. The standard cubic foot is defined at 60°F and 14.73 psi absolute pressure. We will use the ideal gas law to compute the energy per mole for natural gas. The temperature T needs to be converted to Rankin and we need the constant R for the ideal gas law equation.

$$T = 60 + 459.67 = 519.67°R \tag{6.7}$$

$$R = \frac{0.023659 \, lb \, ft^3}{in^2 \, mole \, °R} \tag{6.8}$$

$$PV = nRT \tag{6.9}$$

Substituting into Eq. 6.7:

$$(14.73 \, psi)(1 \, ft^3) = n\left(\frac{0.023659 \, lb \, ft^3}{in^2 \, mole \, °R}\right)(519.67°R) \tag{6.10}$$

Solving, n = 1.198 moles per standard cubic foot of gas. Since the natural gas has 950–1050 BTU per standard cubic foot and there are 1.198 moles in a standard cubic foot, the fuel has 793 to 876 BTU of energy per mole. One mole of methane has a mass of 16 g, so dividing by 16 we get the energy density in BTU per gram for the fuel.

$$\text{Energy Density of Natural Gas} = 49.56 \text{ to } 54.78 \, BTU / gram \tag{6.11}$$

We also know that 16 g of methane will produce 44 g of carbon dioxide and 36 g of water. By multiplying by (16/44) and (16/36) respectively the following can be obtained:

- 18.02 to 19.92 BTU energy for each gram of CO_2 produced
- 22.03 to 24.35 BTU energy for each gram of H_2O produced

We can combine this information with the information on pollution discussed previously and make the following conclusions about the combustion of natural gas. For carbon dioxide assume on average we get 19 BTU energy for each gram of CO_2 produced and for water assume 23 BTU energy per gram produced.

$$\frac{1 \times 10^9 \, \text{BTU}}{\dfrac{19 \, \text{BTU}}{\text{gram} \, CO_2}} = 5.263 \times 10^7 \, \text{g} \, CO_2 = 116{,}000 \, \text{lb} \, CO_2 \tag{6.12}$$

$$\frac{1 \times 10^9 \, \text{BTU}}{\dfrac{23 \, \text{BTU}}{\text{gram} \, H_2O}} = 4.349 \times 10^7 \, \text{g} \, H_2O = 95{,}600 \, \text{lb} \, H_2O \tag{6.13}$$

We burn fossil fuel for the energy. We might use the energy in a power plant to produce electricity, or we might use the energy to heat our homes. There are many industrial and commercial uses for natural gas. The best way to quantify pollution and carbon emissions of a fossil fuel is in terms of a billion BTU of energy. We will compare the results for natural gas to the pollution and carbon emissions for gasoline, diesel fuel, and coal on the basis of a billion BTU of energy. This is how we will assess which fuels are "cleaner" or "dirtier". Cleaner fuels will create less pollution and carbon emissions in providing us with a billion BTU of energy. In summary, for one billion BTU of energy from natural gas we get [1]:

- 40 lb carbon monoxide (CO)
- 1 lb Sulfur dioxide (SO_2)
- 92 lb Nitrous Oxide (NO_x)
- 7 lb particulate
- 116,000 lb carbon dioxide (CO_2)
- 95,600 lb water (H_2O)

The numbers above are good average values for natural gas. Natural gas is produced from many different regions of the country and the sulfur dioxide and particulate pollution will vary significantly depending on the source. Assuming a good combustion process, the estimates for carbon monoxide, nitrous oxide, carbon dioxide and water are independent of the source of the natural gas.

In examining the results above it should be clear that most combustion products from burning natural gas are carbon dioxide and water. Most of the pollution generated comes from the other four products, though there are trace amounts of many different types of pollution. One billion BTUs is about 300 megawatt-hours of energy. A power plant is about 33% efficient in converting the energy into electricity, so a billion BTU is a typical value of natural gas to be used in one day in a natural gas power plant. Looking at the numbers above gives an estimate of what is produced in a day at a typical natural gas power plant.

6.4 **Combustion of Diesel Fuel**

Diesel fuel is a hydrocarbon fuel made from a blend of hydrocarbon molecules. We will study an average sized hydrocarbon molecule for diesel fuel to get a representative chemical reaction. Diesel has an average of 23 hydrogen atoms for each 12 carbon atoms. It is a blend of sizes. In order to draw a sensible hydrocarbon molecule there must be an even number of hydrogen atoms, so it is not possible to draw a sensible molecule for $C_{12}H_{23}$. In order to illustrate typical hydrocarbon molecules that may be included in the diesel fuel blend, consider the $C_{10}H_{20}$ molecule in **Figure 6.8** below:

FIGURE 6.8   Typical $C_{10}H_{20}$ molecule.

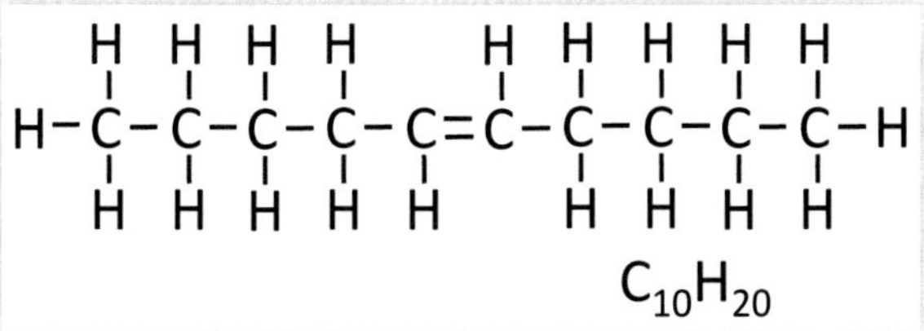

When drawing sensible molecules, each carbon atom needs to have 4 bonds and each hydrogen atom has one bond. Two carbon atoms can form a double bond as illustrated near the center of the $C_{10}H_{20}$ molecule. The double bond doesn't need to be in the middle as shown; it could be between any two carbon atoms. If there were two double bonds in the carbon chain it would be a $C_{10}H_{18}$ molecule, and there may be $C_{10}H_{18}$ molecules in the blend that makes up diesel fuel.

The carbon chains can be branched, in addition to having double bonds. To illustrate possible branching consider the $C_{15}H_{28}$ molecule illustrated in **Figure 6.9** below.

FIGURE 6.9   Typical $C_{15}H_{28}$ molecule.

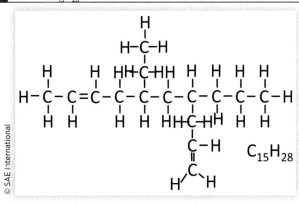

In all of the hydrocarbon molecules please recognize that the four carbon bonds are at the corners of a tetrahedron rather than in a plane as drawn in the figures. The molecules are more bent and twisted in a three-dimensional sense than can be illustrated in two dimensions. For complete combustion all the molecules can combine with oxygen to form carbon dioxide and water as the combustion products. There is a blend of sizes of molecules in the diesel fuel such that the average molecule has a chemical composition of $C_{12}H_{23}$, even though there are no molecules in the mix that have exactly that chemical composition. We can do the chemistry assuming a $C_{12}H_{23}$ molecule, and it gives the right answer even though the actual chemistry is more complex. Complete combustion yields:

$$4C_{12}H_{23} + 71O_2 \rightarrow 48CO_2 + 46H_2O \tag{6.14}$$

We use air to provide the oxygen, which means there will also be nitrogen in the combustion chamber, and there will be nitrous oxides formed during combustion. Incomplete combustion results in small amounts of carbon monoxide being generated. Diesel fuel also contains small amounts of hydrogen sulfide which will be converted to sulfur dioxide. The next step in the analysis is to compute the molecular weights of the components in the chemical reaction.

$$\text{Molecular weight of } 4C_{12}H_{23} = 4\left[(12)(12) + (23)(1)\right] = 668\,g \tag{6.15}$$

$$\text{Molecular weight of } 48CO_2 = 48\left[12 + (2)(16)\right] = 2112\,g \tag{6.16}$$

$$\text{Molecular weight of } 46H_2O = 46\left[(1)(2) + 16\right] = 828\,g \tag{6.17}$$

The basic chemical analysis shows that 668 g of diesel fuel will produce 2112 g of carbon dioxide and 828 g of water. Experimental measurements show that diesel fuel has 138,700 BTU of energy per gallon and that the density of diesel fuel is 832 g per liter. Recognizing there are 0.26417 gal in one liter, it can be shown that the energy density of diesel fuel is:

$$\text{Energy Density Diesel Fuel} = \left(\frac{138,700\,\text{BTU}}{\text{gal}}\right)\left(\frac{0.26417\,\text{gal}}{\text{liter}}\right)\left(\frac{\text{liter}}{832\,g}\right) = 44.04\,\text{BTU}/g \tag{6.18}$$

Since we get 2112 g carbon dioxide and 828 g water for each 668 g of diesel fuel burned, it follows that:

$$\left(\frac{44.04\,\text{BTU}}{g\,\text{diesel}}\right)\left(\frac{668\,g\,\text{diesel}}{2112\,g\,CO_2}\right) = \frac{13.93\,\text{BTU}}{g\,CO_2} \tag{6.19}$$

$$\left(\frac{44.04\,\text{BTU}}{g\,\text{diesel}}\right)\left(\frac{668\,g\,\text{diesel}}{828\,g\,H_2O}\right) = \frac{35.5\,\text{BTU}}{g\,H_2O} \tag{6.20}$$

Eq. 6.20 is included to complete the analysis. The water vapor going into the atmosphere is not regarded as a problem. The carbon dioxide is what scientists believe is contributing to global warming. Equation 6.19 is the more important result. Looking back at the analysis for natural gas it was shown that natural gas gives us 18.02–19.92 BTU of energy per g CO_2 generated. Comparing the two fuels, the natural gas gives us more energy for each gram of carbon dioxide generated. Therefore, we say that natural gas has a lower carbon footprint than diesel fuel. We get more energy out of the natural gas for the same amount of carbon dioxide generated. The next fuel to be studied is coal, and it will be shown that coal has a larger carbon footprint than diesel fuel, and a much larger carbon footprint than natural gas.

In addition to the carbon dioxide, diesel vehicles emit NO_X compounds, carbon monoxide, and particulate. The particulate or soot from diesel vehicles contains carcinogens and used to be much more significant than it is today. Improvements in diesel fuel allow diesel engines to emit far lest particulate than they did 20 years ago. Sulfur has all but been eliminated for the diesel fuel we use today. In order to be able to compare to natural gas, I've used data collected in 2018 [6] and converted it into pounds per billion BTU of heat energy in the fuel. For each billion BTU energy in the fuel, the following emissions will be generated:

- 85 lb NO_X [Compared to 92 lb for natural gas]
- 1480 lb CO [Compared to 40 lb for natural gas]
- Zero pounds SO_2 [Compared to 1 lb for natural gas]
- 2.2 lb Particulate [Compared to 7 lb for natural gas]
- 158,265 lb CO_2 [Compared to 116,000 lb for natural gas]
- 62,050 lb H_2O [Compared to 95,600 lb for natural gas]

These numbers are typical for diesel powered cars and trucks, and for ocean vehicles when they are near port. Ships at sea are still allowed to use a diesel fuel that has a higher sulfur content (and is less expensive). When at sea, ships will emit more SO_2 than what is shown above, but the other emissions are similar. There are policies in process to reduce the SO_2 emissions for ships at sea, and these emissions will probably be reduced significantly over the next 20 years.

The chemistry associated with burning gasoline is assigned as one of the homework problems. Gasoline is a cleaner fuel in the sense that it produces essentially zero particulate and sulfur dioxide emissions. For comparison, NO_X emissions are approximately 127 lb/billion BTU and carbon monoxide emissions are approximately 1740 lb/billion BTU [7]. Gasoline engines emit high amounts of carbon monoxide compared to diesel engines. Carbon dioxide emissions for gasoline are slightly less than for diesel fuel, but there is not a lot of difference. In the grand scheme, diesel, gasoline, and jet fuel all have about the same carbon emissions. Natural gas has significantly lower carbon emissions and coal has significantly higher carbon emissions than the fuels used by vehicles.

In the homework you will calculate the carbon dioxide concentration in the atmosphere and compare to the carbon emissions of the USA and of the world. Please follow the instructions in the homework. You will find that the world emissions of carbon dioxide from burning fossil fuels more than account for the increase in carbon dioxide in the atmosphere. Some of the carbon dioxide generated from burning fossil fuels is being absorbed by the oceans. You will also study the carbon dioxide emissions of cars in the homework and calculate the grams of CO_2 per mile traveled. In the literature you will find the carbon emissions of cars given in grams of CO_2 generated per mi traveled (grams CO_2/mi). If you know the mpg rating for the vehicle it is a straightforward calculation. Suppose a small, diesel-powered car gets 46 mpg. Diesel has 138,700 BTU/gallon and in Eq. 6.19 we calculated that the fuel provides 13.93 BTU of energy for each gram of CO_2 generated. The calculation is as follows:

$$\left(\frac{138,700\,\text{BTU}}{\text{gal}}\right)\left(\frac{\text{gal}}{46\,\text{mi}}\right)\left(\frac{g\,CO_2}{13.93\,\text{BTU}}\right)=216.5\,g\,CO_2\,/\,\text{mi} \qquad (6.21)$$

Once you have worked through the chemistry for gasoline you will be able to make the calculations for gasoline powered cars too. For electric cars we will use the information from the Department of Energy that says in the USA we generated 12.6 Quads of electric energy in 2017. The EPA estimates that 1730 million metric tons of carbon dioxide was generated in producing this electricity in 2017. From these values you can calculate the number of grams of CO_2 generated for each kW-h of electric energy (469 g CO_2/kW-h). Then if an electric car can travel 100 miles on 40 KWH of electric energy in the batteries the carbon footprint is calculated as:

$$\left(\frac{469\,g\,CO_2}{\text{kW-h}}\right)\left(\frac{40\,\text{kW-h}}{100\,\text{mi}}\right)=187.6\,g\,CO_2\,/\text{mi} \qquad (6.22)$$

As you go through the calculations you will find that, for vehicles of similar size, the hybrid-electric and fully-electric vehicles have similar carbon footprint. The regular gasoline powered cars have a significantly higher carbon footprint than electric cars or hybrid electric cars of similar sizes. As long as we produce the majority of our electricity using fossil fuels there is not a huge environmental benefit in driving electric cars. If we can use cleaner sources and/or greatly reduce the amount of electricity generated using fossil fuels, then the electric cars are better for the environment. But it is currently impractical to power large trucks, ships, and aircraft using electric power. It would require a massive change in infrastructure to power trains with electricity, but powering trains with electricity is possible. Japan uses electricity to power almost all of their trains. Going to electric cars can reduce the carbon emissions and pollution generated by cars, and at this point in time transitioning everyone to driving electric cars seems like a realistic possibility. It will be more difficult to electrify the large trucks, ships and aircraft because they travel long distances and require large amounts of energy to make the trip. The size and weight of the batteries will reduce the amount of cargo or passengers the vehicles can carry. Electrifying large trucks and ships seems more possible than aircraft. As discussed in Unit 1, 40% of the weight of a fully loaded Boeing 747 is jet fuel. All the batteries we have today are at least an order of magnitude heavier than jet fuel for the amount of stored energy. At this point in time, it is not possible to make a battery powered aircraft to replace our jets. There is consideration of using liquid hydrogen as the fuel, which would have no carbon footprint, but which would be very expensive.

6.5 **Combustion of Coal**

Coal is complicated, because the composition depends on where it is mined. There is a wide variety of composition in coal. Coal also contains a significant amount of non-burnable material (mostly dirt and rock). The softer coals contain a significant amount of water, which shows up as hydrogen and oxygen in the spectrographic analysis of the coal. Coal can be divided into five basic categories as:

1. Lignite is the softest coal that is used for generating power.
2. Sub-Bituminous coal is a little harder, and is the coal mined in the Powder River Basin in Wyoming and burned in many power plants.
3. Bituminous coal is a little harder yet. Eastern coal used in many power plants is bituminous.
4. Steam coal is a little harder than Bituminous and softer than Anthracite.
5. Anthracite coal is a little too hard to be ground up and used in power plants.

A spectrographic analysis can be performed on coal samples to determine the weight fractions of the different elements in the coal. Coal will have significant amounts of carbon, oxygen, hydrogen, and sulfur. There will be trace amounts of many other elements. For the purpose of combustion analysis, we will only consider the carbon, oxygen, hydrogen, and sulfur. For generating electricity, we are primarily interested in bituminous coal, though significant amounts of lignite, and sub-bituminous are used. **Table 6.1** gives reasonable ranges of values for these types of coal [8].

The elemental percentages in Table 6.1 are in terms of mass. There is a lot of variation in coal depending on where it is mined and different authors may give values that are slightly different from what is given in the table, but these are good average values. Some coal has small amounts of nitrogen, which I have not included in the table. In order to explain the combustion of coal we need reasonable values to work with, and the values in Table 6.1 will work well in helping the reader understand the combustion of coal.

The oxygen in the coal is tied with hydrogen as water. The energy in the coal comes from burning the carbon and the hydrogen that is not tied to the oxygen as water. Burning the sulfur also contributes a small amount of energy. If there is nitrogen in the coal, it contributes almost nothing to the chemical reaction or the energy content of the coal.

The primary pollutants associated with burning coal are carbon monoxide, sulfur dioxide, nitrous oxides, and particulate. These are the same pollutants produced by burning other fossil fuels, but burning coal generates a lot more particulate than other fossil fuels. Coal is a solid fuel that is mined and there is a significant amount of dirt,

TABLE 6.1 Typical chemical content and energy density of coal.

Coal Type	Carbon %	Hydrogen %	Oxygen %	Sulfur %	Energy (BTU/lb)
Lignite	60–75	6.0	33.5–16	0.5-3	7300–10,700
Sub-Bituminous	75–82	5.9	18.1–11.1	1	10,400–11,700
Bituminous	82–88	5.8-5.6	11.2–5.4	1	11,700–12,800

minerals, and moisture mixed with the coal. Companies have shaker operations to shake the dirt and minerals off and will wash the coal with water to remove more of the dirt and minerals. The coal is dried to remove most of the moisture. But it is inevitable that the coal delivered to power plants will contain a significant amount of dirt, minerals, and moisture. The dirt and minerals will be ground up with the coal and go through the combustion process in the furnace, but they will not burn. This is what generates the coal ash, or particulate, that comes from burning coal. As a rule of thumb you can estimate that about 5% of the coal delivered to the power plant will become coal ash (or particulate). If a large coal plant uses 20 train-car loads of coal per day, it will generate about one train-car load of coal ash per day. Modern coal plants have technology to remove almost all the particulate and prevent it from going up the smokestack. One hundred years ago, a large fraction of the particulate went up the smokestack and became air pollution. In addition to breathing the particulate, people who lived near the power plant would wake up every morning to a thin layer of black goo deposited on buildings, roads, and everything else. It was a very dirty environment. Modern coal plants emit very little particulate to the atmosphere. Coal plants are much cleaner than they were 100 years ago, but they generate a lot of coal ash which must be landfilled. Coal ash contains trace amounts of heavy metals and carcinogens and we need to be careful with disposal to prevent causing a health hazard.

We need a way to estimate the pollution from burning coal so we can compare to the pollution from burning natural gas and other fossil fuels. The values for sulfur dioxide and particulate are going to vary a lot depending on where the coal is mined. The values in **Table 6.2** below are approximate. In the table, "Brown Coal" refers to lignite and sub-bituminous coal similar to what is mined in the Powder River basin. "Hard Coal" is bituminous and is similar to eastern coal. Please use Table 6.2 to estimate the pollution generated in burning coal. The numbers in the table are pounds of pollution for each billion BTU of heat energy in the coal [9].

Most of the particulate is removed at the power plant and is landfilled as coal ash. The other pollutants will go into the atmosphere. If you compare to the values for diesel fuel and natural gas, the pollution from coal is higher in all cases. Burning gasoline produces more carbon monoxide than coal, but gasoline is lower for all other emissions. Particulate emission is much higher in coal than in other fuels, and this needs to be discussed.

With liquid and gas fuels it is possible to put them in a large settling tank where most of the heavier impurities will settle to the bottom. It is an inexpensive way to reduce the contamination in the fuel. Gas and liquid fuels can also be run through a filter, further removing contamination. These simple inexpensive processes of settling and filtering remove nearly all the contamination and make gas and liquid fuels cleaner.

TABLE 6.2 Typical pollution values for hard and brown coal.

Pollutant	Hard Coal (lb)	Brown Coal (lb)
SO_2 (lb/billion BTU)	1779	3166
NO_x (lb/billion BTU)	679	426
CO (lb/billion BTU)	207	207
Particulate (lb/billion BTU)	2798	7569

© SAE International

Coal is a solid material and it is much more difficult to remove contamination from a solid. There are mechanical processes to remove the dirt and minerals that inevitably come mixed with the coal. The industry does a good job of removing impurities. Only a few percent of the coal delivered is non-combustible material, but power plants use a lot of coal, so a few percent end up being a lot of material. Coal plants remove almost all the particulate; very little is put into the atmosphere. Trace amounts of mercury and radon will be emitted as gasses because they cannot be removed. The particulate matter becomes the coal ash that must be disposed. The coal ash is largely dirt and minerals that have been ground into a fine powder and run through the furnace, with trace amounts of heavy metals and carcinogens. Breathing the dust is hazardous to our health, but properly landfilled coal ash does not present a large environmental problem. There is a steady stream of news articles about power companies that did not properly landfill their coal ash, causing environmental problems. From a practical viewpoint, some of the coal ash will cause environmental problems.

Samples of coal are studied using a mass spectrometer to determine the mass concentration of the elements in the coal. Averaging several samples gives an average composition for the coal. The results of the analysis can be used to study the chemistry of the coal. As an example, assume that the analysis indicates the coal is 80% carbon, 5.9% hydrogen, 13.1% oxygen and 1% sulfur by mass. The chemistry is done by number of atoms rather than by mass so the first step in the analysis is to divide each mass percentage by the atomic mass of the element and add them up. As with previous analysis I have rounded the masses to 12 for carbon, 1 for hydrogen, 16 for oxygen and 32 for Sulfur.

$$\frac{80}{12} + \frac{5.9}{1} + \frac{13.1}{16} + \frac{1}{32} = 13.4167 \qquad (6.23)$$

The atomic percentages are then obtained by dividing the appropriate term on the left side by the sum on the right side of the equation. For example, the carbon percentage is:

$$Carbon\% = \frac{\dfrac{80}{12}}{13.4167} \times 100\% = 49.6894\% \qquad (6.24)$$

The percentage by number of atoms in the coal is then:

- Carbon = 49.6894%
- Hydrogen = 43.9752%
- Oxygen = 6.1025%
- Sulfur = 0.2329%

The atoms are combined in the coal mostly as hydrocarbons, water and hydrogen sulfide. The oxygen content indicates that there is significant moisture in the coal, and some of the hydrogen is tied to oxygen atoms as water. Some of the hydrogen is tied with carbon forming hydrocarbons. Some of the carbon exists as pure carbon graphite. The sulfur is probably tied with either hydrogen or oxygen as sulfur dioxide or hydrogen sulfide. The mass spectrometer analysis tells us the atomic percentages in the coal, but it cannot tell us how the atoms are tied together as molecules. When doing the chemistry of coal, we do not need to know exactly how the atoms are tied together. We will start with the percentage compositions above and write a balanced equation. This method of chemical analysis yields a good approximation to the burning of coal. The atomic components in the coal will be combined with oxygen in the air to form carbon dioxide, water and sulfur dioxide. The left side of the chemical equation is written as:

$$(49.6894)C + (43.9752)H + (6.1025)O + (0.2329)S + (\underline{\hspace{1.5cm}})O_2 \rightarrow \qquad (6.25)$$

At this point in the analysis, we do not know how much oxygen will need to be added for complete combustion. All the carbon must become carbon dioxide. All the hydrogen must become water, and all of the sulfur must become sulfur dioxide. The right side of the chemical equation is written as:

$$(49.6894)CO_2 + \left(\frac{43.9752}{2}\right)H_2O + (0.2329)SO_2 \qquad (6.26)$$

The oxygen required for combustion will be provided from air. There is a blank that goes with the oxygen in the air in Eq. 6.25 that will be calculated. On the right side all the carbon will form carbon dioxide, so the number with the carbon dioxide is 49.6894. The hydrogen will all form water, so the number with the water is half of 43.9752. All the sulfur will form sulfur dioxide, so the number with the sulfur dioxide is 0.2329. The oxygen on the left side from the air is then calculated so that the total oxygen on the left and right sides balance. Total oxygen on the right side is:

$$(49.6894)(2) + (21.9876) + (0.2329)(2) = 121.832 \qquad (6.27)$$

Subtract the 6.1025 oxygen already on the left side in the coal and then divide by two to get the O_2 that comes from air. The final balanced equation is:

$$(49.6894)C + (43.9752)H + (6.1025)O + (0.2329)S + (57.8649)O_2 \rightarrow$$
$$(49.6894)CO_2 + (21.9876)H_2O + (0.2329)SO_2 \qquad (6.28)$$

The total molar weight of the coal on the left side is:

$$(49.6894)(12) + (43.9752)(1) + (6.1025)(16) + (0.2329)(32) = 733.341\,g\,Coal \quad (6.29)$$

The molar weights for the carbon dioxide, water, and Sulfur dioxide on the right side are:

$$(49.6894)\left[12 + (2)(16)\right] = 2186.33\,g\,CO_2 \qquad (6.30)$$

$$(21.9876)\left[(1)(2) + 16\right] = 395.777\,g\,H_2O \qquad (6.31)$$

$$(0.2329)\left[32 + (2)(16)\right] = 14.906\,g\,SO_2 \qquad (6.32)$$

The conclusion from the chemistry is that burning 733.341 g coal will generate 2186.33 g CO_2, 395.777 g H_2O, and 14.906 g SO_2. This process is helpful in determining the amount of carbon dioxide and sulfur dioxide generated in burning coal.

Another lab test must be performed to determine the heat energy that comes from burning the coal. For the example coal, assume the heat energy is 11,200 BTU/lb. That is, burning 1 lb of the coal will generate 11,200 BTU of heat energy. Recognizing that

733.341 g coal generates 2186.33 g CO_2, and that there are 453.6 g/lb, the carbon footprint of the fuel can be calculated as:

$$\left(\frac{11,200\,\text{BTU}}{\text{lb Coal}}\right)\left(\frac{\text{lb}}{453.6\,\text{g}}\right)\left(\frac{733.341\,\text{g Coal}}{2186.33\,\text{g CO}_2}\right) = 8.28\,\text{BTU}\,/\,\text{g CO}_2 \qquad (6.33)$$

The water and sulfur dioxide can be estimated in a similar manner. Water is not usually regarded as a problem as far as emissions. The estimate for sulfur dioxide is:

$$\left(\frac{11,200\,\text{BTU}}{\text{lb Coal}}\right)\left(\frac{\text{lb}}{453.6\,\text{g}}\right)\left(\frac{733.341\,\text{g Coal}}{14.906\,\text{g SO}_2}\right) = 1214.8\,\text{BTU}\,/\,\text{g SO}_2 \qquad (6.34)$$

At this point we have discussed three fossil fuels, natural gas, diesel fuel, and coal. With natural gas we get 18.02 to 19.92 BTU of heat energy for each g CO_2 generated. An average value for natural gas is 18.97 BTU per g CO_2. Diesel fuel provides 13.93 BTU per g CO_2, and the coal example provides 8.28 BTU per g CO_2. It was discussed earlier that a typical value for a power plant was to use a billion BTU of energy in one day. Based on using a billion BTU of energy, a natural gas plant would produce 116,000 lb of CO_2. A diesel plant would produce 158,000 lb of CO_2 and a coal plant using the example coal would produce 266,000 lb of CO_2. For a typical hard coal it is possible to estimate that 1814 lb of SO_2 would also be generated in providing a billion BTU of energy. **Table 6.3** below provides a summary of the three fossil fuels studied and the emissions produced.

TABLE 6.3 Carbon dioxide and pollution emissions per billion BTU energy.

Emission	Natural Gas	Diesel	Hard Coal	Soft Coal
CO_2 (lb)	116,000	158,000	266,000	245,000
SO_2 (lb)	1	Zero	1779	3166
NO_x (lb)	92	85	679	426
CO (lb)	40	1480	207	207
Particulate (lb)	7	2.2	2798	7569

© SAE International

From Table 6.3 it should be clear that natural gas is the cleanest of the fossil fuels considered. Coal is the dirtiest of the fossil fuels. Gasoline was assigned as a homework problem. The carbon dioxide emissions for gasoline are similar to that of diesel fuel. Gasoline has essentially zero sulfur dioxide and particulate emissions. The carbon monoxide and nitrous oxide emissions are 1740 lb and 127 lb per billion BTU. The carbon monoxide emissions for gasoline and diesel fuel are relatively high because they are used in internal combustion engines. If gasoline and diesel were used to fire the burner for a steam turbine the carbon monoxide emissions would be much lower. Comparisons are based on a billion BTU, which is a typical amount of energy used in a medium sized natural gas power plant on an average day. Large coal plants use several billion BTU of energy on a typical day, and a diesel-powered generating station would use much less than a billion BTU per day. Assuming a 33% thermal efficiency in the power plant, using a billion BTU would produce 97.7 MW-h of electric energy.

6.5.1 **Carbon Emissions for Common Items**

Most of the energy we use comes from fossil fuels. Virtually everything we purchase was partially manufactured and transported using fossil fuel. Everything we purchase has a "carbon footprint" because fossil fuels are used in the manufacturing and transportation. There is a group of scientists who have tried to quantify how much carbon dioxide is generated from different products. The information is posted at the web site http://www.co2list.org/. I included this information for the readers who are interested in how to calculate the carbon footprint of common everyday items. The web site is maintained by a group of scientists who want to reduce carbon footprint. It is a summary of many data sources and attempts to include everything in estimating the total carbon footprint.

The carbon footprint for natural gas and coal on the above web site is higher than what we calculated in the previous section. Our calculations include only the carbon generated from burning the fuel. The web site includes the carbon associated with mining and transporting the fuel to where it is burned in addition to the carbon associated with burning the fuel.

I include the table below for curiosity. There is more information at the web site for those who are interested [10] (**Table 6.4**).

TABLE 6.4 Carbon dioxide to produce common items.

Food
22 kilos CO_2 per kilo of red meat
5.9 kilos CO_2 per kilo of chicken, fish or eggs
2.9 kilos CO_2 per kilo of cereal or carbohydrate
1.7 kilos CO_2 per kilo of fruit or vegetables
Miscellaneous
381 kilos CO_2 per square meter, to build houses
225 kilos CO_2 per square meter, to make solar cells
3.2 kilos CO_2 per liter of heating oil
2.4 kilos CO_2 per cubic meter of natural gas
0.8 kilos CO_2 per kwh of electricity
0.5 kilos CO_2 per dollar spent in the USA, average
Transportation (These include making vehicle, road, rails, airports, etc.)
0.47 kilos CO_2 per passenger-km on flights of 800 km
0.35 kilos CO_2 per passenger-km on flight of 1,300 km
0.37 kilos CO_2 per passenger-km on cruise
0.34 kilos CO_2 per kilometer in a 12 kpl car (28.2 mpg)
0.28 kilos CO_2 per kilometer in a 17 kpl car (37.6 mpg)
0.11 kilos CO_2 per passenger-km in a train
0.13 kilos CO_2 per kilometer bicycling on bike lane
0.05 kilos CO_2 per kilometer bicycling without bike lane
0.20 kilos CO_2 per passenger-km in local bus
0.05 kilos CO_2 per passenger-km in long distance bus

6.6 **Homework**

Partial correct answers are included to give you a way to check your answers.

1. The mass of the earth's atmosphere is approximately 5.15×10^{18} kg. Assume that this corresponds to approximately 1.78×10^{20} moles of gas molecules. Our current estimate is that the carbon dioxide content of the atmosphere is about 410 ppm, which means that for every million gas molecules in the atmosphere, approximately 410 of them are carbon dioxide. The carbon dioxide content of the atmosphere has been increasing about 2.2 ppm each year for the past 15 years.

    For this homework problem I want you to make an estimate of the significance of the carbon footprint of the United States.

    a. Based on 410 ppm, calculate the number of moles of carbon dioxide in the atmosphere. Then calculate the total number of g CO_2 in the atmosphere.

    b. Based on an increase of 2.2 ppm annually, calculate the number of moles of CO_2 that are added to the atmosphere each year. Then calculate the total number of g of CO_2 added to the atmosphere.

    c. Based on data from the EPA the USA emitted 5270.7 million metric tons of CO_2 in 2017. How many g of CO_2 does the United States add to the atmosphere each year? [Answer: 5.2707×1015 g, or about 0.673 ppm in the atmosphere.]

    d. Compare the US CO_2 emissions each year to the 400-ppm total CO_2 in the atmosphere and the 2.2 ppm increase in CO_2 each year. Write a short paragraph about the comparison and explain whether you feel that US carbon emissions are significant or insignificant as far as building CO_2 in the atmosphere.

2. In 2018 the world CO_2 emissions were estimated as 32,915.9-million metric tons. This estimate is a little low because it only includes CO_2 emissions from the burning of fossil fuels. Most CO_2 emissions come from burning fossil fuel, so it is a low but reasonable estimate.

    a. With this estimate, how many grams of CO_2 does the world add to the atmosphere each year?

    b. Compare the world CO_2 emissions each year to the 410-ppm total CO_2 in the atmosphere and the 2.2 ppm increase in CO_2 each year. Write a short paragraph about the comparison and explain whether you think the world carbon emissions are significant or insignificant as far as building CO_2 in the atmosphere. [Answer: Adds 4.2 ppm to the atmosphere each year.]

3. The Nissan Leaf is a fully electric car capable of traveling 107 miles on its 30 KWH battery system. Electric cars are regarded socially as having zero carbon footprint, but a significant portion of the electricity used in the cars will come from fossil fuels. In this problem you are to estimate the carbon footprint of the Nissan Leaf in grams of CO_2 per mile traveled (grams CO_2/mile).

In the US in 2018, we produced 12.9 Quads of electric energy according to the information from Lawrence Livermore National Laboratory. The EPA estimate is that we generated 1732-million metric tons of carbon dioxide in 2017 generating electricity. For this problem assume that we will generate 1732 million metric tons of carbon dioxide to generate 12.9 Quads of electric energy. From this information you should be able to calculate the grams of carbon dioxide generated per kW-h of electric power (grams CO_2/KWH). Then knowing that the Nissan Leaf can go 107 miles on 30 kW-h electric energy you should be able to calculate the grams per mile CO_2 generated by the Nissan Leaf. [Answer: The United States generates 458 grams of CO_2 for each kilowatt-hour of electric energy. *For those who are interested, in 2015 the USA generated 553 grams CO_2 per kilowatt-hour. The carbon emissions have declined because of the increase in renewable energy used to generate electricity and because of the increase in natural gas used to generate electricity and the decline in the use of coal.*]

Charging a battery is not 100% efficient. It will require approximately 33 kW-h of electricity to slow charge the batteries to 30 kW-h of useful energy. Fast charging of the batteries will require more than 33 kW-h. Please account for it taking 33 kW-h to charge the batteries in your calculations. [Answer: 141.3 g CO_2 per mile traveled.]

4. The Toyota Prius is a hybrid electric car that is rated at 52 mpg on regular gas. Assume the gasoline has 125,000 BTU per gallon. Calculate the carbon footprint of the Toyota Prius in gCO_2 per mile. Compare this to the Nissan Leaf. [You will need to work problem 7 before doing this problem to find the carbon footprint of gasoline.]

5a. The Honda Fit is a small gasoline car rated at 32 mpg. Calculate the carbon dioxide footprint of the car in grams per mile. [Answer: 281 g CO_2 per mile] Compare to the Nissan Leaf and Toyota Prius.

5b. A 1971 Oldsmobile 98 had a fuel economy of about 8 mpg. Gasoline only cost about $0.20 per gallon in those days. Calculate the carbon dioxide foot-print of a car that gets 8 mpg.

6. You have calculated the carbon footprint for three small hatchback cars. The Nissan Leaf is a fully electric car with a cost of $35,455.00 for a mid-priced model. The Toyota Prius is a hybrid car with a cost of $28,650.00 for a mid-priced model. The Honda Fit is a gasoline car with a cost of $17,900.00 for a mid-priced model.

   a. Assume that electricity has a cost of $0.12 per kW-h and that gasoline has a cost of $2.15 per gallon. Calculate the energy cost per year for the three cars assuming they are driven 15,000 miles per year.

   b. Consider the costs of purchasing and owning the vehicles and the carbon emissions associated with the vehicles. Write a couple of paragraphs telling me which vehicle you think is the best purchase and why. There is no right or wrong answer for this question. Please tell me what you think is the best overall choice.

7. Assume that the typical gasoline molecule has the formula C_8H_{18}.

   a. Draw the molecule.

   b. Work out the chemical reaction assuming complete combustion.

   c. Calculate the number of BTUs of energy in the gasoline for each gram of CO_2 generated. Assume gasoline has 125,000 BTU per gallon. Assume that the density of gasoline is 770 g/l. [Answer: 13.89 BTU per gram CO_2]

   d. Compare the carbon footprint of gasoline to those of natural gas, diesel and coal.

8. Suppose a power plant is generating electricity using coal. Assume that coal is being used is 80% C, 6%H, 13%O, and 1% S by weight. The energy density of the coal is 26,100 kJ/kg = 11,220 BTU/lb.

   a. Work out the chemical reaction for the coal.

   b. Calculate the number of BTUs energy from the coal for each gram of carbon dioxide generated.

   c. Compare to the other fuels discussed in class.

9. Assume that we generate 225 kg of CO_2 in manufacturing one square meter of solar panels, as is listed on the www.co2list.org web page. In Springfield, Missouri, we estimate that one square meter of solar panel will produce 1000 KWH of electric energy each year, and that the solar panels will last for 20 years. Calculate the carbon footprint in grams per KWH for the solar panels. [Answer: 11.25 g CO_2 per kW-h. For comparison power plant produce 468 g CO_2 per kW-h.]

References

1. https://en.wikipedia.org/wiki/Natural_gas#Environmental_effects
2. https://en.wikipedia.org/wiki/Snowball_Earth
3. https://en.wikipedia.org/wiki/Milankovitch_cycles
4. https://climate.nasa.gov/vital-signs/sea-level/
5. https://climate.nasa.gov/vital-signs/carbon-dioxide/
6. https://www.bts.gov/content/estimated-national-average-vehicle-emissions-rates-vehicle-vehicle-type-using-gasoline-and
7. https://www.bts.gov/content/estimated-national-average-vehicle-emissions-rates-vehicle-vehicle-type-using-gasoline-and
8. https://en.wikipedia.org/wiki/Coal_assay
9. https://en.wikipedia.org/wiki/Coal
10. http://www.co2list.org/.
11. Scheffler, M., Feyrer, K., and Matthias, K., *Fördermaschinen—Hebezeuge, Aufzüge, Flurförderzeuge (Hoisting machines—Hoists, elevators, industrial trucks)* (Braunschweig/Wiesbaden: Friedr. Vieweg & Sohn Verlagsgesellschaft mbH, 1998), ISBN:3-528-06626-1.

The Electric Power Grid

7.1 Introduction

If we drive electric cars, the electricity to power them will come from the electric power grid. It is important that you have a basic understanding of how we generate and distribute electric power. **Figure 7.1** illustrates hourly and seasonal variation in the demand for electric power in the mid-Atlantic region of the USA in 2009. In Figure 7.1, PJM stands for "Pennsylvania-Jersey-Maryland," the region of the country used in generating the figure. The region includes significant portions of the three states. The graph starts on Monday. Time zero on the graph is midnight separating Sunday and Monday [1].

For the mid-Atlantic region, the highest use of electric energy is in the summer months because electricity is used for air conditioning in most homes and many buildings. The lowest use is in the spring and fall, and winter months are in-between. Most people and businesses in the mid-Atlantic region have access to natural gas for heating, which makes the electric loads in winter less than in summer. In regions of the country where most of the customers use electricity for heat, the winter months will show more demand for electricity than the summer months. Electricity usage is larger during the week than on the weekend. The trend on this graph is typical for many parts of the country and world. Demand for electricity is lowest in the middle of the night and highest in the afternoons and early evenings. Power companies must provide the electricity we need when we need it, and our need for electricity varies with time of day, day of the week, and seasonally.

Figure 7.1 illustrates that there is a daily cycle of how we use electricity. There is a large difference between how much electric power we use in the middle of the night and in the

FIGURE 7.1   Average hourly load.

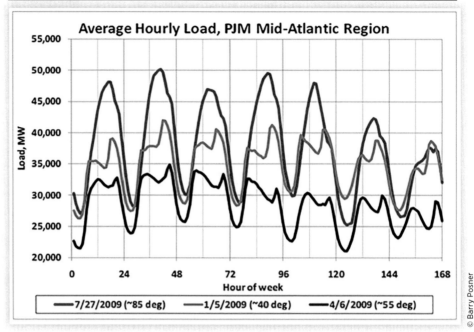

afternoons and evenings. There is a large difference seasonally too. In most parts of the world we use a lot of electric power in the summer to provide air conditioning, and in the winter to provide heat. We use lower amounts of electric power in the spring and fall when temperatures are moderate. The electric power industry must provide power when we need it, and they do a good job. We are seldom without electric power in the USA. Our electric power system is so reliable that we take it for granted and it causes us hardship when there is a large weather event that knocks the power out. We are very fortunate. In many parts of the world the electric power system is not reliable.

7.1.1 Why Electricity?

A question that sometimes arises is why do we use so much electricity? Electricity is not a naturally occurring source of energy. We use primary sources of energy to produce electricity. It would be more efficient to use the primary sources of energy directly. The answer is that electricity can be transmitted over large distances and distributed to homes and businesses with very little energy loss. That's what makes electricity valuable. We can take any primary source of energy, convert it into electricity, and use the grid to distribute the energy to customers. When we use electricity, we use a blend of the primary energy sources that were used to make the electricity: wind, solar, natural gas, coal, etc. Electricity is only as "clean and green" as the sources used to create the electricity.

Power companies must always provide exactly the amount of electric power that people want to use. If demand exceeds supply, we will have a blackout. If supply exceeds demand, the voltage will rise and begin destroying the power system. Balancing the power is largely automated. Computers control the power output at the power plants and make sure the grid is balanced. There are people in control rooms monitoring the power system making sure that supply and demand are always balanced. These people control when power plants go on-line and off-line. By using weather forecasts and historical records of the electricity demand in the past they can plan for how much power needs to be produced during the day. The overall goal is to produce the required electric power from the least expensive sources available, but there are many other considerations. In practice, balancing the grid is rather complex.

There is protection circuitry that will shut the system down and protect the power grid if supply-and-demand gets out of balance. We cannot destroy the power grid. The grid system will shut down and protect itself if the power gets out of balance. Ideally, we would have a way to store large quantities of electric energy so that we could store excess electric energy when the demand is low, and provide the energy later at a time when demand is high. But we do not have a good way to store large quantities of electric energy. Electricity must be generated as it is needed to keep the system in balance. There is a great deal of research being done to develop ways of storing large quantities of energy, which will be discussed later. If we are successful it will greatly change how we balance the power grid. The discussion below is the current methods used to balance the electric power grid.

7.1.2 Base Load Power

Base load power is the power that must be generated all the time. Base load power is the minimum power in Figure 7.1 that must be generated each day in the middle of the night. For base load power we want to use power plants that have a low fuel cost per kilowatt-hour to generate the electricity, since they will be running all the time. Base load power plants tend to have a low fuel cost and a high capital cost.

7.1.3 Peak Load Power

Peak load power is the power necessary to handle the higher loads during the day. Power plants are brought on-line during the day and turned off at night. Fuel cost is less important for the peak load power plants because they are not operating all the time. Capital cost is more important for the peak load power plants.

Electric power is bought and sold all the time to keep supply and demand in balance. Power companies can produce power from their power plants, or they can buy power from another power company. Power generated in the middle of the night is less valuable than power generated during the day. Overall, base load power is less valuable per kW-h than peak power. In most of the country power companies are required by law to charge individual households the same cost per kW-h regardless of when we use the power, but business and industry will pay a different price for the electricity depending on when they want to use it. In 2002, Texas became more deregulated than the rest of the country, and household customers can get real time market prices in Texas. The buying and selling of electricity in Texas is significantly different from the rest of the country because of the deregulation.

Wind and solar energy have a zero-fuel cost. All the costs for wind and solar power are from the capital costs of the equipment and the maintenance costs. We cannot control how much power we get from wind and solar energy, which can cause problems in balancing the grid. We can use weather forecasts to estimate how much wind and solar power we will get at different times of the day, but it is not possible to control the wind or the sun or clouds. We currently use other power plants, mostly natural gas power plants, to keep the grid in balance. The way we currently use wind and solar power is to get all the energy we can get from the wind and sun and use other power plants to balance the load. Wind and solar can be used for a portion of the base load power but are not reliable enough to provide all the base load power.

The cost and subsidies associated with the wind and solar power that has been installed are such that we need to use all the energy they can generate. The subsidy is based on the number of kW-h of energy produced and shutting the wind turbine down means losing the federal subsidy, which is not a good financial decision. We do not want to turn them off. The subsidy is set to end in 2020, and, as I write this book, I do not know whether it will be renewed or not. Local power companies have indicated that the cost of wind and solar has declined to the point that they plan to install more wind and solar power even if the subsidies end. Their current cost estimates for generating electricity from wind power make it less expensive than any other energy source. As wind and solar become less expensive, and as the subsidies end, it will make sense to use wind and solar to help balance the power grid. We can install a large excess of wind and solar generation capacity and then turn off what we do not need. It is similar to what we do with natural gas peaking plants today. It does not make sense financially to do that with the wind and solar installed to date. But if the cost of electricity generated from wind and solar continues to decline, and if the subsidies end, it will make sense to do that in the future.

Nuclear power has a very high capital cost and a low fuel cost. Nuclear power is used as base load power. Ideally, we run the nuclear power plant at full capacity all the time to minimize the overall cost of the electricity generated. We can control the power generated in the nuclear power plant and reduce or increase power when necessary to balance the grid.

Coal power has a relatively high capital cost and a low fuel cost. Coal provides base load power. We reduce the power from the coal plants at night, but do not shut them down. It takes quite a while to get a coal plant started once it has been shutdown. The capital cost of a coal plant is much lower than for a nuclear power plant. The fuel cost for a coal plant is higher than a nuclear plant.

Natural gas power plants are the primary way we handle the peak loads that happen during the day. They have a lower capital cost and higher fuel cost than coal power plants. Natural gas power plants can be shut down and restarted each day. With fracking, the cost of natural gas has declined in recent years and the overall cost of producing electricity from natural gas is comparable to coal, or in many places less expensive than coal. In the current market, building a natural gas plant makes more sense economically than building a coal plant.

Hydro power is generally used to assist with peak load power. Hydro is low cost electricity to produce and it generates essentially no pollution or carbon dioxide. Environmentalists complain that building the lake destroys natural habitat, but overall hydro power is the cleanest and greenest energy source. There are a limited number of places where it makes sense to build a lake and dam. We do not have unlimited hydro power available. On most of the days there is a limited amount of water that can be drawn from the lake. We must maintain a suitable lake level. Power companies will use the hydro power available everyday when they feel they can get the most value from it, which is generally during peak hours.

Oil power plants and diesel generators provide a very small percentage of the electric power we use but are important in generating the peak power during the summer months, when usage is high or for emergencies. These power plants have a low capital cost and a very high fuel cost. Some of these power plants may only be used a few times each year, so the capital cost is most of the cost of operating them.

Fuel prices are variable. I will summarize fuel prices today as I write this book, but the numbers I give will become inaccurate with time. A good source for current fuel prices is the US Energy Information Administration [2]: https://www.eia.gov/state/seds/data.php?incfile=/state/seds/sep_prices/total/pr_tot_US.html&sid=US.

Producing electricity from fossil fuels requires heat energy (BTU) to fire the burner, generate the steam, and power the steam turbine. The heat energy can come from any energy source. The information below is to be used to compare the fuel cost for the commonly used fuels.

- Coal: There is considerable variation in the cost of coal depending on location. The lowest cost listed was $29.15 per short ton (2000 lb) in Iowa and the highest was $99.00 per short ton in New Hampshire. The average cost in the USA was $30.09 per short ton. The BTU content varies depending on where the coal was mined. The average for coal in the USA is about 19,270,000 BTU per short ton. Thus, the average cost for a million BTU is $30.09/19.27 = $1.56 per million BTU.
- Natural Gas: Power plants purchase natural gas for less than what homeowners pay. The average cost for power plants in the USA is currently $3.68 for 1000 ft^3. On average natural gas has 1030 BTU per cubic foot, so the cost of the natural gas for power plants is $3.57 for a million BTU. Average residential cost for natural gas is about $10.52 per thousand cubic feet.
- Propane: For wholesale pricing the average in 2018 was $0.965 per gallon. It requires 10.9285 gallons to make one million BTU, so the cost for a million BTU is $10.55. Residential propane in 2018 was $2.483 per gallon, which is $27.14 per million BTU.
- Gasoline: The USA average cost in 2018 was $2.813 per gallon. There are 121,000 BTU per gallon, so this translates to a cost of $23.25 for a million BTU.
- Diesel: The USA average cost in 2018 was $3.178 per gallon. There are 138,700 BTU per gallon, so this translates to a cost of $22.91 for a million BTU.
- Nuclear: People working in the power industry say that nuclear fuel is the lowest for the amount of energy, and I feel confident the statement is true. However, reliable data are unavailable. I believe the cost of a million BTU of energy from nuclear fuel will be less than the cost of coal.
- Wind and Sun: There is no fuel cost associated with wind and solar energy. The wind and sun are free.

The fuel cost for coal is significantly less than for natural gas, but the capital cost of the power plant is much less for natural gas than for a coal plant. The operational and maintenance cost for a natural gas plant is less than a coal plant. The coal plants are shutting down, as the cost of the electricity generated by the coal plants is too expensive compared to power industry. Natural gas is relatively expensive today. Historically the electricity produced in coal plants has been less expensive than electricity produced in natural gas plants. Today natural gas and wind power are both less expensive than coal, and we see the coal plants being closed. But it is difficult to predict the future. If we reduce the amount of fracking the cost of natural gas will probably increase, and, at some point coal could become less expensive than natural gas. Building a power plant is a 50+ year investment and power companies need to make their decisions on what they think will be best over the next 50+ years.

There is a perpetual myth that propane is as economical as natural gas in heating your house. It is a myth. Natural gas is less expensive and has always been less expensive than heating with propane. Propane is a good economical choice if natural gas is not available.

7.2 **Nuclear Power**

Nuclear power is used to generate 22.1% of the electricity used in the US. The nuclear energy is used to heat water and convert it to steam, and then the steam drives a steam turbine, which drives a generator to make the electric power. The process is similar to coal and natural gas power plants.

The radiation from nuclear fuel is dangerous and hazardous to our health. A large part of the cost of building a nuclear power plant is related to containing and controlling the radiation from the nuclear fuel. Special materials must be used for some components in the nuclear plant because not all materials can withstand the nuclear radiation.

There is an essentially unlimited amount of nuclear fuel that could be used in power plants. The fuel cost for a nuclear power plant is low compared to other fuel sources. The capital cost of building a nuclear power plant is very high. Getting permits and approvals to build a nuclear power plant is difficult. There are always public protests to building a nuclear power plant, and the protests lead to delays and cost over-runs. Nuclear accidents, such as Pennsylvania's Three Mile Island in 1979, Ukraine's Chernobyl in 1986, and Japan's Fukushima in 2011, have had a large impact turning public opinion against nuclear power. In the early years of the Obama presidency plans were made to build new nuclear power plants in the USA, but the Fukushima accident caused many of those plans to be delayed or canceled. We completed the Watts Bar Unit 2 in Tennessee that began operation in 2016. This was the first new nuclear power plant to begin operation in 20 years. The Watts Bar Unit 1 plant in Tennessee, which began operation in 1996, is the second newest nuclear power plant in the US. There are a few nuclear power plants that are scheduled to be completed in the next decade, but it remains to be seen how many will be completed and put into service.

In my opinion the future of nuclear power currently looks bleak. In the US and Europe, nuclear plants are closing faster than they are being built. The high cost of building nuclear plants makes it difficult to find investors. The low cost of electricity produced by natural gas and wind power make the economics of nuclear power questionable. On the positive side we have an almost unlimited amount of nuclear energy and nuclear power does not cause air pollution or global warming. Properly contained nuclear power is a very clean energy source. Disposing of nuclear waste is a difficult issue. There is no site in the US approved for nuclear waste disposal. In the US all the nuclear waste is stored on-site at the nuclear power plants. The biggest challenge for nuclear power is to convince the public that nuclear power is safe. It seems that every time we get to a point where the public is ready to embrace nuclear power there is another accident that turns the public against it. There are many good things about nuclear power, but it cannot be successful without public support.

7.3 **Hydroelectric Power**

Hydroelectric power is an old and successful technology. Water wheels were used for hundreds of years to provide power for grinding grain or sawing lumber, etc. Modern water turbines are used to capture energy from the flowing water and produce electric power. Hydroelectric power is one of the greenest sources of energy. There are carbon emissions associated with building the dam and lake, but very little carbon or pollution emissions in operating the hydroelectric power plant. Building a lake destroys habitat and covers private property. There are always protests and public concern when building the lake.

Hydropower is also low in cost. If there were more suitable places to build hydroelectric dams we would probably build more of them. The energy associated with hydroelectric power is gravitational energy, mass × gravity × height, as taught in physics classes. By creating a lake behind the dam we have a large mass of water at a height above the water turbine. The illustration below is helpful in illustrating the energy available from the lake (**Figure 7.2**).

If V is the usable volume of water in the lake and ρ is the density of the water, then the mass of usable water in the lake is ρV. The volume V will be the surface area of the lake multiplied by the depth of water that can be removed and run through the water turbines. The water turbines will be about 90% efficient in converting the energy into electricity. Putting this together the following equation can be developed.

$$\text{Hydro}\left(\text{Gravitational}\right)\text{Energy} = \rho V g h = \gamma V h \qquad (7.1)$$

FIGURE 7.2   Schematic of hydro power.

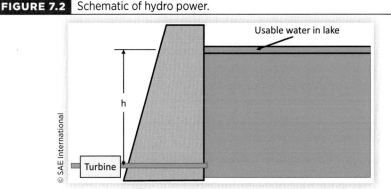

Where γ is the weight density of the water. The area of a lake is usually measured in acres or square miles. Drawing one foot of water from a lake is a large volume and mass of water and represents a large amount of energy. The amount of water that can be drawn from the lake varies with rainfall and evaporation, which to some extent varies according to time of year. There are times when the water must be removed from the lake faster than it can flow through the water turbines. The flood gates are opened, and energy is dumped down the spillways. There will be extremely dry times when we will not be able to draw any water from the lake. The power company will try to maximize the value of the electricity produced, using the hydro energy to produce electricity when it is most valuable.

Example 7.1: Table Rock Lake in Missouri covers ~43,100 acres [3]. How many MW-h of electric energy could be produced by lowering the lake level 6 in? Assume that the weight density of water is 62.4 lb/ft³, that the water turbine is 90% efficient, and that the height of the water above the turbine is 185 ft.

An acre is 43,500 ft². The volume of water in 6 in depth can be calculated as:

$$\text{Volume} = \left(43{,}100 \text{ acre}\right)\left(\frac{43{,}500 \text{ ft}^2}{\text{acre}}\right)\left(\frac{6}{12}\text{ ft}\right) = 9.3872 \times 10^8 \text{ ft}^3 \qquad (7.2)$$

The gravitational energy is the volume of water multiplied by the 185 ft height above the water turbines. One MW-h of electric energy is 2.655×10^9 ft-lb. Since the water turbines will be about 90% efficient in converting the gravitational energy to electricity, the electric energy is calculated as:

$$\text{Electric Energy} = \left(0.90\right)\left(9.3872 \times 10^8 \, ft^3\right)\left(\frac{62.4 \text{ lb}}{ft^3}\right)\left(185 \, ft\right)\left(\frac{\text{MW-h}}{2.655 \times 10^9 \text{ ft-lb}}\right)$$

$$= 3673 \text{ MW-h}$$

$$(7.3)$$

To put this in perspective the power station at Table Rock Lake has four 50 MW generators, a total of 200 MW capacity. The energy in 6 in of water in the lake would be enough to power the generators for 18.4 h. On most days there is less than 6 in of usable water in the lake, and the generators will operate shorter periods of time. Also please understand that we do not reduce the level of the lake 6 in in drawing the water out. There is water constantly flowing into the lake and we need to balance maintaining a fairly-steady water level with drawing water out of the lake to make electric power.

The dam and lake serve more purposes than generating electricity. Justification for building the dam and lake will be based on flood control, which is the most important reason for building the dam and lake. The lake will provide a source of recreation that many people will enjoy. The lake provides a reliable source of fresh water. If we divide the cost of the lake and dam by all the uses and benefits the cost of generating electricity from hydropower is very low. Even if we charge the whole cost of building the lake and dam to generating electricity the cost of generating the electricity is 3 - 5 cents per kilowatt-hour. Electricity generated using hydropower is inexpensive compared to other methods of generating electricity.

There are negative aspects of hydropower too. When building a hydroelectric power plant, we must build a dam and lake. The property for the lake must be acquired. Some owners will sell willingly, and some property will need to be acquired by eminent domain. Environmentalists will be concerned that the lake destroys natural habitat. Building the dam will reduce the flow of water in the river below the dam, which will have negative impacts on the people who live below the dam. The people below the dam will be concerned about failure of the dam. More damage and death are caused by dam failure than any other electric power plants, except perhaps nuclear plants. There will need to be public hearings and political activity to obtain the land and permits to build the dam and lake.

7.4 **Wind Power**

Wind power was a vital part of transportation for hundreds of years. Wind powered the ships. Europeans colonized the world using wind power. Developing powerful sailing naval ships allowed European nations to become dominant world powers. Sailing vessels averaged about 5 mph on the oceans. The ships traveled faster when the wind was blowing, and slower when it was not blowing.

As we developed the steam engine during the 1800 s, sailing ships were replaced by steam ships. The steam engines were more expensive than sails. Steam engines were noisy and dangerous and required fuel and lots of maintenance. Boilers exploded and ships burned. There were lots of problems with the early steam engines but with all their faults, they were an improvement over the sailing ships. Steam power was faster and more reliable.

The wind is free and there is always a temptation to use a free energy source. Wind turbines have been used for hundreds of years to pump water. One of the fundamental problems in using wind is that it is not possible to control the wind. You must be opportunistic and use the wind when it is available. Pumping water from a well into a tank or pond is a good use of wind power because you do not need for it to be pumping all the time. In the Netherlands, wind power was used to drain the fields by pumping water from low areas into the ocean. Wind turbines are not a new idea, but they are becoming important in generating electricity.

The electric grid must be balanced. We must generate exactly the amount of power that we are using. Wind power is subsidized by the US government (Production Tax Credit) at 2.3¢ per kW-h, for each kW-h produced by the wind turbines. The PTC is scheduled to be phased out at the end of 2020, but it is possible that it will be extended. The owners of wind turbines would like for them to produce as much electricity as possible because the subsidy they receive from the government is proportional to the amount of electric energy produced. That approach has worked fine as long as wind power is a small percentage of the electricity produced. As wind power becomes a larger percentage of the electricity generated, we will need to be able to turn the wind turbines down at night to reduce the amount of energy they generate. It is technically possible to turn the wind turbines down or off now. The turbines are all designed so that we can twist the blades and cause the rotation to slow down or stop, but this makes them less profitable for the owners. The owners will lose the 2.3¢ per kW-h subsidy for each kW-h the wind turbine could have produced, but we must balance the electric grid. As wind power becomes a larger percentage of the electricity we use, it will be necessary to reduce power from the wind turbines at night.

We use a peak amount of electric power during the day and into the early evenings. We must have a way to handle these peak loads. If we have excess wind power available from wind turbines that are shut off, then we can bring them online to generate the power we need. This requires a large capital investment in wind turbines that are turned off most of the time. Most of the cost associated with wind power is capital cost. If the wind turbines are off most of the time, the number of kW-h produced each year will be much less than if they are on all the time. This makes the cost of the electricity much more expensive because the annual capital depreciation cost and other annual costs are divided by a smaller number of kW-h of electricity. It becomes a matter of cost. If wind turbines become inexpensive enough, we can afford to have a lot of excess wind power and use it to balance the grid and handle peak loading.

There are other ways to balance the grid. Solar power is available during the day, and in the future it may be possible to handle a large part of the daytime peak loads using solar power. In the near future we will continue to use natural gas power plants to handle most of the peak loads, along with hydropower and a small amount of oil-fired power plants and diesel generators. Coal and nuclear will be used as baseline power in the near future but are currently more expensive than using wind or natural gas. With the current costs, we will continue to see less electricity generated using coal and nuclear because they are the more expensive solution. However, it is hard to predict the future. If natural gas and wind become more expensive in the future, we may see a resurgence of coal and nuclear power.

In an ideal situation we would store some of the excess electric energy generated at night and use it during the day when we need more electricity. We have been experimenting with this for more than 100 years. It is not a new idea. The problem has always been that we do not have a good economical way to store large quantities of electric energy. One of the most successful experiments has been using excess electric power to pump water into a lake, and then using water turbines to extract the energy from the lake when we need it. In Missouri, the Taum Sauk project was built in the early 1960s using this idea. In the middle of the night, electricity produced at the power plant has almost no value, so the electricity is used to pump water into a lake. During the day when the electricity has more value, the water in the lake is run through water turbines and used to make electricity. There is a large loss of energy due to pumping and turbine efficiencies and evaporation from the lake. Lots of energy maybe stored in a lake, but with the pumping losses and evaporation, we will get less than half of the energy back as electricity. It is not very efficient, but it is cost effective and it is a way we can store large quantities of energy.

Another approach to storing excess electric energy generated at night is to store it in batteries. The batteries can then be used to handle the peak loads during the day.

In Unit 1, we studied electric cars and learned that the batteries for electric cars must be lightweight for the amount of energy they store. Weight is not as critical for utilities. Utility batteries will be stationary, so the main design considerations will be cost and reliability. Storing enough energy to power a city would require a huge number of batteries. There are no battery systems currently available that can realistically store the amount of energy needed to power a city. There are multiple utilities around the country trying pilot projects with batteries storing small amounts of electric power. Utility batteries are not practical today, but they could become practical in the future. Another possibility is that if most people drove electric cars, and if we allow the electric company to control when the car batteries are charged, and if the electric company can draw charge from the batteries when they need it, and if, if, if… It might be possible to use a large number of electric cars as the battery storage medium for electric energy.

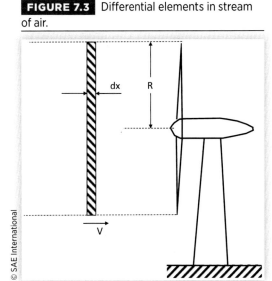

FIGURE 7.3 Differential elements in stream of air.

© SAE International

The energy associated with wind is kinetic energy. There is a stream of air moving through the wind turbine blades. The cross-sectional area of the stream of wind is πR^2, where R is the turbine radius. Assume there is a mass of air of thickness dx moving toward the wind turbine at speed V. **Figure 7.3** illustrates the differential element of air moving toward the wind turbine.

Recognizing that kinetic energy is $\tfrac{1}{2}mV^2$ we can write the mass and energy of the differential mass of air as shown below:

$$\text{Differential Mass} = \rho\pi R^2 dx \qquad (7.4)$$

$$\text{Differential Energy} = \frac{1}{2}\left(\rho\pi R^2 dx\right)V^2 \qquad (7.5)$$

The power of the wind approaching the wind turbine is the derivative of energy with respect to time. Dividing Eq. 7.5 by dt and recognizing that dx/dt is the velocity V of the wind it follows:

$$\text{Power} = \frac{1}{2}\rho\pi R^2 V^3 \qquad (7.6)$$

Equation 7.6 is the power in the wind stream approaching the turbine blades. The turbine will not be able to absorb all of the energy in the wind, because absorbing all of the energy in the wind would require that the turbine stop the wind, i.e., bring the velocity of the wind to zero as it hits the turbine. This is not physically possible, because the wind must flow through the turbine blades. The turbine blades will slow the speed of the wind as it flows through but not stop it. Equation 7.6 will greatly overestimate the amount of power the wind turbine can produce. We will need a better model for doing the analysis of wind turbines.

Wind turbines are similar to propellers except they absorb power from the moving fluid rather than adding power to the moving fluid. The model we will use was originally developed by William Froude. For the turbine to slow the speed of the wind stream, the conservation of mass law in fluid mechanics requires that the radius of the incoming

FIGURE 7.4   Wind turbine model.

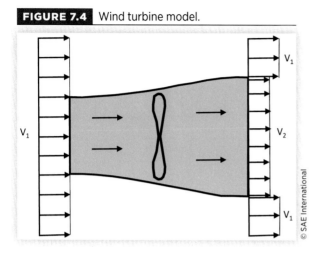

© SAE International

air stream must be smaller than the existing air stream behind the turbine. **Figure 7.4** is similar to the figure for developing the propeller model.

In the analysis V_1 is the speed of the wind and V_2 is the speed of the wind after it has been slowed by the wind turbine. If R is the radius of the turbine blades then the radius of the air stream approaching the turbine must be smaller than R and the radius of the air stream behind the turbine must be greater than R. Conservation of mass requires that $A_1V_1 = A_2V_2$ where A_1 is the area of the stream of wind approaching the turbine and A_2 is the area of the stream of wind behind the turbine. Since $V_1 > V_2$ it must be true that $A_2 > A_1$. Just as with the propeller derivation, Froude assumed the velocity V of the wind at the turbine is the average of V_1 and V_2.

$$V = \frac{V_1 + V_2}{2} \tag{7.7}$$

With this assumption, the conservation of momentum formula can be used to find the force acting on the turbine:

$$\text{Force} = \frac{1}{2}\rho\pi R^2 \left(V_1^2 - V_2^2\right) \tag{7.8}$$

The power that the wind turbine absorbs from the stream of wind is the product of the force and the velocity V of the air at the turbine.

$$\text{Power} = \frac{1}{2}\rho\pi R^2 \left(V_1^2 - V_2^2\right)V = \frac{1}{2}\rho\pi R^2 \left(V_1^2 - V_2^2\right)\left(\frac{V_1 + V_2}{2}\right) \tag{7.9}$$

The theoretical efficiency of the turbine e_{turb} is given by Eq. 7.10 below:

$$e_{turb} = \frac{1}{2}\left[1 - \frac{V_2^2}{V_1^2}\right]\left(1 + \frac{V_2}{V_1}\right) \tag{7.10}$$

The above formulas assume an optimum design for the turbine blades. The maximum theoretical efficiency is 59.3%. That is, the maximum amount of power that can be absorbed by the wind turbine is 59.3% of the energy in the wind stream approaching the wind turbine, due to the nature of how it works. In reality the efficiency will be 40%–50% for most wind turbines.

Eq. 7.10 can be used to prove the 59.3% maximum possible efficiency. Substitute $x = V_2/V_1$ so that the efficiency is only a function of x. Take the derivative with respect to x and set it equal to zero to find the x values where the efficiency is at a minimums and maximums. The solution will yield x = 1/3 for the maximum. Back substitute into the efficiency equation and the value is 59.3%.

The goal for the information on wind turbines is to give the reader a way to estimate the energy and power that is possible based on the specifications of the wind turbine. Companies sometimes give optimistic estimates of the energy that can be recovered from the wind turbine. The reader should be able to perform analysis to estimate what is possible for a well-designed wind turbine.

Example 7.2: The company Nature Power makes a home wind turbine priced at $2500.00 and rated at 2000 W in a 28-mph wind. The specifications say it will produce 350 kW-h per month in a 12-mph wind. The rotor diameter is 6 feet. Assume that the density of air is 1.22 kg/m³. In addition to buying the wind turbine it will cost about $2000.00 to have a 30-ft tall tower erected for the wind turbine.

1. For a first approximation assume the ideal turbine efficiency and use the analysis in this class for wind turbines. How much power will the turbine produce in a 28-mph wind?

    Solution: I will work the solutions in the metric system. The wind speed V_1 = 28 mph, which is equal to 12.51 m/s. The diameter is 6 ft = 1.828 m, so the radius R = 0.9144 m. Assuming the ideal efficiency of the turbine means that V_2/V_1 = 1/3, so V_2 = 12.51/3 = 4.17 m/s. Plugging everything into the power equation (Eq. 7.9):

$$\text{Power} = \frac{1}{2}(1.22)\pi\left(0.9144^2\right)\left(12.51^2 - 4.17^2\right)\left(\frac{12.51 + 4.17}{2}\right) = 1861\,\text{W} \quad (7.11)$$

    The analysis shows that a 6-ft diameter turbine under the most ideal conditions can produce 1861 watts in a 28-mph wind. This assumes a perfectly designed turbine and that the generator and inverter are 100% efficient, which is not possible or realistic. In reality, the wind turbine will produce significantly less than 1861 watts. The manufacturer rates the wind turbine at 2000 W in a 28 mph wind. This turbine will not perform as well as advertised by the manufacturer.

2. As a more realistic set of assumptions, assume that the turbine is 50% efficient and that the generator and inverter are 90% efficient. Under these assumptions how much power will the turbine produce in a 28-mph wind?

    Solution: When the turbine efficiency is given, start with the efficiency formula (Eq. 7.10) and solve for the V_2/V_1 ratio. You will probably need to use the root finder in a calculator, and it is convenient to substitute x = V_2/V_1 into the formula in using the calculator. The equation is:

$$0.50 = \frac{1}{2}\left[1 - x^2\right](1 + x) \quad (7.12)$$

    Solving this equation yields x = 0.618, and since we know V_1 = 12.51 m/s it follows that V_2 = (0.618) (12.51) = 7.73 m/s. In analyzing wind turbines, it is common to start with the wind speed (V_1) and the efficiency of the turbine. When this is the case start with the efficiency formula (Eq. 7.12) and solve for x and the velocity of the air behind the wind turbine (V_2). The turbine power is then calculated using the power formula:

$$\text{Turbine Power} = \frac{1}{2}(1.22)\pi\left(0.9144^2\right)\left(12.51^2 - 7.73^2\right)\left(\frac{12.51 + 7.73}{2}\right) = 1570\,\text{W}$$

$$(7.13)$$

The turbine will provide 1570 W power on the shaft that drives the generator. If the generator and inverter are 90% efficient in converting the turbine power into electric power, it follows that the electric power generated is:

$$\text{Electric Power} = (0.90)(1570) = 1413 \text{ W} \tag{7.14}$$

This is far less than the 2000 W promised by the manufacturer.

3. The manufacturer also indicates that the wind turbine produces 350 kW-h of energy each month if the wind speed averages 12 mph. How much energy per month would the wind turbine produce in a 12-mph wind? Assume 50% turbine efficiency and that the generator and inverter are 90% efficient. Converting the 12 mph wind speed to metric units it follows that $V_1 = 5.363$ m/s, and since the efficiency of the turbine is 50% $V_2 = (0.618)(5.363) = 3.315$ m/s. The turbine power and electric power is:

$$\text{Turbine Power} = \frac{1}{2}(1.22)\pi(0.9144^2)(5.363^2 - 3.315^2)\left(\frac{5.363 + 3.315}{2}\right) = 123.6 \text{ W} \tag{7.15}$$

$$\text{Electric Power} = (0.90)(123.6) = 111.2 \text{ W} \tag{7.16}$$

Assuming that the wind turbine produces 111.2 W of continuous power for a 30-day month it follows that the electric energy produced by the turbine in a month is:

$$\text{Electric Energy} = (111.2 \text{ W})\left(\frac{30 \text{ day}}{\text{month}}\right)\left(\frac{24 \text{ h}}{\text{day}}\right)\left(\frac{\text{kW}}{1000 \text{ W}}\right) = 80 \text{ kW} - \text{h} / \text{month} \tag{7.17}$$

This is much less than the 350 kW-h specified by the manufacturer. This wind turbine will not be able to produce 350 kW-h per month in a 12-mph wind. It is possible to estimate the average wind speed required to produce the 350 kW-h per month. First work backwards through Eq. 7.17 to find that the average power required is 486 W. That is, the wind turbine would need to provide 486 W average power for 30 days to generate 350 kW-h of electric energy. The generator and inverter are 90% efficient and for a 50% efficient turbine the V_2/V_1 ratio is 0.618. Substituting all of this into the power equation yields the following:

$$\frac{486 \text{ W}}{0.90} = \frac{1}{2}(1.22)\pi(0.9144^2)\left(V_1^2 - ((0.618)V_1)^2\right)\left(\frac{V_1 + (0.618)V_1}{2}\right) \tag{7.18}$$

You will probably need to use the root finder on your calculator to solve the equation. The solution is $V_1 = 8.768$ m/s = 19.6 mph. The conclusion is that the average wind speed must be 19.6 mph or greater for the wind turbine to produce 350 kW-h of electric energy each month.

4. The wind turbine will cost $4500.00 to install. The next step is to decide if the wind turbine is a good financial investment. For this analysis assume that the wind turbine will produce 80 kW-h of electric energy per month and that the value of the electricity is $0.12 per kW-h. Most household consumers in the Mid-West pay less than $0.12 per kW-h, but that is a typical value for the west coast. The monthly value would be ($0.12)(80) = $9.60, and the yearly value would be (12)($9.60) = $115.20, which is a 2.56% return on investment. CDs are paying less than 2.5% interest. Assuming no maintenance cost on the wind turbine it would be a better investment than a CD. Investing in mutual funds or bonds will give a better return on the investment compared with purchasing the wind turbine.

5. The last step for this example is to use Excel to plot power output of the wind turbine as a function of wind speed up to 35 mph. Assume 50% turbine efficiency and 90% efficiency for the generator and inverter as in other parts of the example problem. Assume that the power output is zero at speeds below 7 mph. **Table 7.1** is the first several lines of the spreadsheet used to calculate the graph in **Figure 7.5**. Figure 7.5 illustrates how wind speed effects power output. The power output from a wind turbine varies with the cube of wind speed, so a small change in wind speed causes a large change in the power output.

TABLE 7.1 Spreadsheet for Example 7.2.

Prop. Dia.	1.8288	Meters			V_2/V_1	0.618	0.500
Efficiency	0.50						
Air Density	1.22	Kg/m³					
						Corrected	
mph	m/s	m/s	m/s	Watts	Watts	Watts	
Wind speed	V_1 speed	V_2 speed	V speed	Turbine	Turbine	Electricity	
0	0	0	0	0	0	0	
1	0.446944	0.276212	0.361578	0.071532	0	0	
2	0.893889	0.552423	0.723156	0.57226	0	0	
3	1.340833	0.828635	1.084734	1.931376	0	0	
4	1.787778	1.104847	1.446312	4.578078	0	0	
5	2.234722	1.381058	1.80789	8.941558	0	0	
6	2.681667	1.65727	2.169468	15.45101	0	0	
7	3.128611	1.933482	2.531046	24.53563	24.53563	22.08207	
8	3.575556	2.209693	2.892624	36.62462	36.62462	32.96216	
9	4.0225	2.485905	3.254203	52.14716	52.14716	46.93245	
10	4.469444	2.762117	3.615781	71.53246	71.53246	64.37922	
11	4.916389	3.038328	3.977359	95.20971	95.20971	85.68874	
12	5.363333	3.31454	4.338937	123.6081	123.6081	111.2473	
13	5.810278	3.590752	4.700515	157.1568	157.1568	141.4411	
14	6.257222	3.866963	5.062093	196.2851	196.2851	176.6566	
15	6.704167	4.143175	5.423671	241.4221	241.4221	217.2799	
16	7.151111	4.419387	5.785249	292.997	292.997	263.6973	
17	7.598056	4.695598	6.146827	351.439	351.439	316.2951	

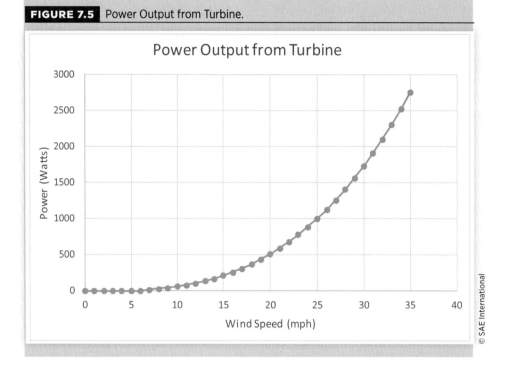

FIGURE 7.5 Power Output from Turbine.

© SAE International

Example 7.3: This example analyzes a much larger wind turbine typical for a utility to purchase and use for providing wind power to their customers. The wind turbine below has a rotor diameter of 100 m and an efficiency of 50% in a 12 m/s wind. Assume the density of air is 1.22 kg/m³. Commercial wind turbines have rotor diameters that are typically 100 m in diameter and cost ~$4,000,000 installed. The goal for this example is to evaluate the power output and the value of the electricity produced.

Solution: We are given the efficiency of the propeller, so use the efficiency formula to get the V_2/V_1 ratio and the value of V_2.

$$0.50 = \frac{1}{2}\left[1-x^2\right](1+x) \tag{7.19}$$

The solution is $x = V_2/V_1 = 0.618$ and $V_2 = (0.618)\,(12\text{ m/s}) = 7.416$ m/s. The thrust force on the turbine is calculated using Eq. 7.8:

$$\text{Thrust Force} = \frac{1}{2}(1.22)\pi\left(50^2\right)\left(12^2 - 7.416^2\right) = 426{,}378\text{ N} = 95{,}620\text{ lb} \tag{7.20}$$

This thrust force creates a large bending moment at the base of the wind turbine. The geotechnical work for the foundation and the structural work for the tower will need to be substantial to carry the large bending moment. The power is calculated using Eq. 7.9 and multiplying by the 90% efficiency.

$$\text{Power} = \frac{1}{2}(1.22)\pi\left(50^2\right)\left(12^2 - 7.416^2\right)\left(\frac{12+7.416}{2}\right)(0.90) = 3{,}725{,}000\text{ W} \tag{7.21}$$

A 12 m/s wind is equal to a 26.85 mph wind. Getting a typical sustained wind that high is uncommon in most of the country. The average wind speed will depend on where the wind turbine is located.

To help illustrate how location impacts the power output of the wind turbine, weather data was collected from weather.com for one day in Springfield, Missouri and is presented in **Table 7.2** below. [Note: Analyzing data for a day is not adequate to decide whether or not to buy a wind turbine, but it illustrates the principle. In practice the reader would want to extend the analysis to a year or more of data in determining the economics of purchasing a wind turbine.] Table 7.2 is the spreadsheet that was used to analyze the wind turbine for the weather data gathered. The wind speed data was put into the V_1 column in the spreadsheet, and the analysis was done using the assumptions above.

In Table 7.2 it was assumed that the wholesale value of the electricity was $0.07 per kW-h. There is a subsidy on wind power of $0.023 per kW-h, so this assumption is that the actual wholesale value is $0.047 per kW-h. The actual value will depend on the time of day and time of year, but this is a reasonable assumption. For a 4-million-dollar

TABLE 7.2 Wind turbine analysis for Springfield, Missouri.

Prop. Dia.	100	meters			$V2/V1$	0.618	0.50			
Efficiency	0.5									
air density	1.22	kg/m³								
Value	$0.07	/kW-h								
Generator	0.90									
Time	V1 (mph)	V1 (m/s)	V2 (m/s)	V (m/s)	Power (W)	Energy (kW-hr)	Value	Total (KWhr)	9584.57	kW-h
Midnight	8	3.5756	2.2097	2.8926	98556	98.56	$6.90	Total ($)	$671	
1:00	8	3.5756	2.2097	2.8926	98556	98.56	$6.90	Monthly ($)	$20,128	
2:00	6	2.6817	1.6573	2.1695	41578	41.58	$2.91	Yearly ($)	$244,886	
3:00	4	1.7878	1.1048	1.4463	12320	12.32	$0.86			
4:00	2	0.8939	0.5524	0.7232	1540	1.54	$0.11			
5:00	2	0.8939	0.5524	0.7232	1540	1.54	$0.11			
6:00	2	0.8939	0.5524	0.7232	1540	1.54	$0.11			
7:00	6	2.6817	1.6573	2.1695	41578	41.58	$2.91			
8:00	8	3.5756	2.2097	2.8926	98556	98.56	$6.90			
9:00	10	4.4694	2.7621	3.6158	192492	192.49	$13.47			
10:00	10	4.4694	2.7621	3.6158	192492	192.49	$13.47			
11:00	10	4.4694	2.7621	3.6158	192492	192.49	$13.47			
Noon	12	5.3633	3.3145	4.3389	332627	332.63	$23.28			
1:00	12	5.3633	3.3145	4.3389	332627	332.63	$23.28			
2:00	14	6.2572	3.8670	5.0621	528199	528.20	$36.97			
3:00	16	7.1511	4.4194	5.7852	788448	788.45	$55.19			
4:00	18	8.0450	4.9718	6.5084	1122615	1122.61	$78.58			
5:00	18	8.0450	4.9718	6.5084	1122615	1122.61	$78.58			
6:00	20	8.9389	5.5242	7.2316	1539938	1539.94	$107.80			
7:00	18	8.0450	4.9718	6.5084	1122615	1122.61	$78.58			
8:00	14	6.2572	3.8670	5.0621	528199	528.20	$36.97			
9:00	14	6.2572	3.8670	5.0621	528199	528.20	$36.97			
10:00	12	5.3633	3.3145	4.3389	332627	332.63	$23.28			
11:00	12	5.3633	3.3145	4.3389	332627	332.63	$23.28			

investment this wind turbine would return about 250-thousand dollars per year in Springfield, Missouri. This is not a good return on the investment. Springfield, like many other places in the country is not a good place for wind power. A second set of data was collected for Sioux Falls, South Dakota, and presented in the spreadsheet below.

This was probably an exceptionally windy day in Sioux Falls. But if this were a typical day the return on investment would be very good. Investing $4,000,000 and getting an annual return of $2,276,668 is an excellent investment. Wind power would make sense in Sioux Falls. To do the analysis correctly the reader should collect wind data for a year or multiple years at a particular location and estimate the return for a typical year. Wind data are available on many weather sites for many locations in the country. It is tedious to obtain the data and run the analysis, but it is a straightforward process.

TABLE 7.3 Wind turbine analysis for Sioux Falls, South Dakota.

Prop. Dia.	100	meters		V2/V1	0.618	0.50			
Efficiency	0.5								
air density	1.22	kg/m³							
Value	$0.07	/kW-h							
Generator	0.90								
Time	V1 (mph)	V1 (m/s)	V2 (m/s)	V (m/s)	Power (W)	Energy (kW-hr)	Value	Total (KWhr)	89106.40	kW-h
Midnight	33	14.7492	9.1150	11.9321	6917594	6917.59	$484.23	Total ($)	$6,237
1:00	33	14.7492	9.1150	11.9321	6917594	6917.59	$484.23	Monthly ($)	$187,123
2:00	32	14.3022	8.8388	11.5705	6307586	6307.59	$441.53	Yearly ($)	$2,276,668
3:00	30	13.4083	8.2864	10.8473	5197291	5197.29	$363.81		
4:00	27	12.0675	7.4577	9.7626	3788825	3788.83	$265.22		
5:00	22	9.8328	6.0767	7.9547	2049658	2049.66	$143.48		
6:00	14	6.2572	3.8670	5.0621	528199	528.20	$36.97		
7:00	14	6.2572	3.8670	5.0621	528199	528.20	$36.97		
8:00	12	5.3633	3.3145	4.3389	332627	332.63	$23.28		
9:00	16	7.1511	4.4194	5.7852	788448	788.45	$55.19		
10:00	18	8.0450	4.9718	6.5084	1122615	1122.61	$78.58		
11:00	23	10.2797	6.3529	8.3163	2342053	2342.05	$163.94		
Noon	24	10.7267	6.6291	8.6779	2661013	2661.01	$186.27		
1:00	26	11.6206	7.1815	9.4010	3383244	3383.24	$236.83		
2:00	26	11.6206	7.1815	9.4010	3383244	3383.24	$236.83		
3:00	30	13.4083	8.2864	10.8473	5197291	5197.29	$363.81		
4:00	32	14.3022	8.8388	11.5705	6307586	6307.59	$441.53		
5:00	33	14.7492	9.1150	11.9321	6917594	6917.59	$484.23		
6:00	32	14.3022	8.8388	11.5705	6307586	6307.59	$441.53		
7:00	30	13.4083	8.2864	10.8473	5197291	5197.29	$363.81		
8:00	28	12.5144	7.7339	10.1242	4225590	4225.59	$295.79		
9:00	26	11.6206	7.1815	9.4010	3383244	3383.24	$236.83		
10:00	24	10.7267	6.6291	8.6779	2661013	2661.01	$186.27		
11:00	24	10.7267	6.6291	8.6779	2661013	2661.01	$186.27		

A problem with wind energy that is difficult to model is the fact that the best places for wind energy are areas that are low in population. When wind turbines are located far from population centers, transmission lines must be constructed to get the energy in to where it will be used and building transmission lines is expensive. The cost of building the transmission lines needs to be part of the capital cost.

7.5 **Solar Power**

Solar power varies with the time of day and the time of year. To analyze solar power there is a standard model used to estimate the intensity of the sunlight on a clear day. The model assumes the earth is spherical and rotates around the sun in a circular idealized orbit. With these assumptions, on a clear day, the intensity of the sunlight and sun angle depend on the latitude, longitude, and time of day. Cloudiness is a major factor in solar intensity and is difficult to predict from weather forecasts. We will add a cloudiness factor to the model, but from a practical viewpoint it is difficult to get an accurate value for the cloudiness factor. Historical data on cloudiness, if available is helpful, but is difficult to get for most locations.

This model neglects many secondary effects. The earth is closest to the sun in early January and furthest from the sun in early July, and this has a secondary impact on the solar intensity. There is less moisture in the air in the winter and early spring, and more moisture in the summer. Moisture in the air absorbs a small amount of solar radiation and reduces solar intensity. The sun emits higher and lower levels of radiation; it is not a constant energy source. In the early morning and late evening, the sunlight must pass through a longer distance of atmosphere because of the low sun angle, which reduces the intensity of the sunlight. The earth is not exactly spherical, and the rotation around the sun is slightly elliptical rather than circular. The earth rotates on a slanted axis as it moves around the sun. All these things are minor secondary effects that are not accounted for in the model. The secondary effects are all small compared to the effect of the position of the sun relative to the earth and the clouds. The model developed in this section is a good approximation for modeling solar energy.

7.5.1 **Angle of Declination**

On the summer solstice (approximately June 22) the sun is directly over the Tropic of Cancer at 23.44° north latitude, and on the winter solstice (approximately December 21) the sun is directly over the Tropic of Capricorn at 23.44° south latitude. For this model we will assume a sinusoidal variation of the declination angle (θ_D) with a small correction factor according to Eq. 7.22 below [4]:

$$\theta_D = -\arcsin\left[0.39779\cos\left(0.98565°(N+10)+1.914°\sin\left(0.98565(N-2)\right)\right)\right] \quad (7.22)$$

Equation 7.22 is an old formula and the angles are given in degrees. It works for finding the declination angle on a calculator, if the calculator is set in degree mode. It is inconvenient programming in Excel, because Excel uses radians for angles. Recognizing

0.98565° is 0.017203 radians and 1.914° is 0.033406 radians, the formula can be modified for Excel as:

$$\theta_D = -\arcsin\left[0.39779\cos\left(0.017203(N+10)+0.033406\sin\left(0.017203(N-2)\right)\right)\right] \quad (7.23)$$

For Eq. 7.22 and Eq. 7.23, N is the day of the year such that January 1 is N = 1. At solar noon, when the sun is the highest in the sky, the angle between a normal to the surface of the earth and the sun is equal to the latitude location of the place on earth minus the declination angle. Example 7.4 illustrates how to calculate the angle of the sun at solar noon.

Example 7.4: Consider a city located at latitude of 37.2° north. Calculate the angle between a normal to the surface of the earth and the sun at solar noon on March 27, 2019. If a solar array were placed parallel to the ground at this location, this would be the angle between the solar array and the sun at solar noon. The intensity of the sunlight striking the solar array is proportional to the cosine of this angle. To do the analysis we find the day of the year for March 27, 2019 and substitute into Eq.7.22 to find the declination angle.

$$N = 31 + 28 + 27 = 86 \quad (7.24)$$

$$\theta_D = 2.62° \quad (7.25)$$

$$\text{Sun Angle} = 37.2° - 2.62° = 34.58° \quad (7.26)$$

On March 27, at latitude of 37.2°, the angle between a normal to the earth surface and the sun at solar noon is 34.58°. The intensity of the sunlight striking the solar array is proportional to the cosine of this angle, which is 0.823, or 82.3% of the direct sunlight. On average, direct sunlight has an intensity of 950 W/m². If the solar array were tilted so it pointed directly at the sun, the intensity of sunlight striking the solar array would be 950 W/m². If the solar array is parallel to the ground, the intensity of the sunlight will be 82.3% of the 950 W/m², which is 782 W/m². The solar cells used for terrestrial applications are typically about 18% efficient in converting sunlight into electricity.

Using excel, it is possible to plot how the declination angle and sun angle vary throughout the year. Continued from the example above the plot was developed assuming that the place on earth where the solar array is located is at latitude 37.2° and that the sun angle is plotted for solar noon (**Figure 7.6**).

Figures 7.7 and **7.8** were developed to help understand how the earth rotates and progresses around the sun. The earth is tilted 23.44° and spinning on an axis as it moves around the sun. The axis is always tilted in the same direction. For the summer solstice (June 21st) the earth is tilted toward the sun as illustrated in Figure 7.7 with the northern hemisphere tilted toward the sun. From the earth it appears as if the sun is directly over the Tropic of Cancer.

FIGURE 7.6   Declination and Sun Angle for 37.2 Latitude

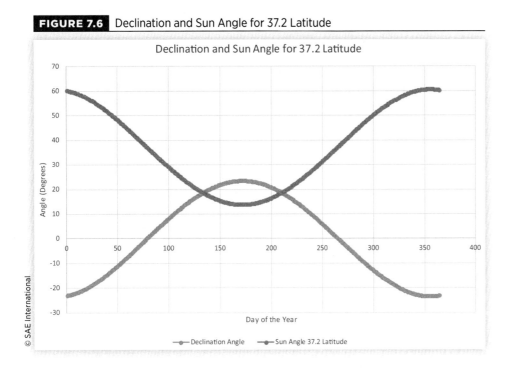

FIGURE 7.7   Summer Solstice.

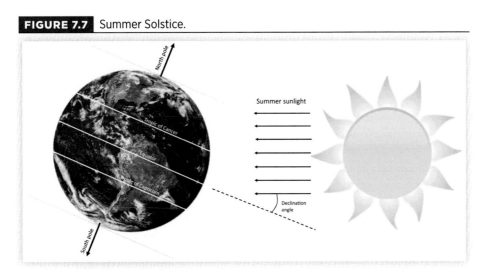

The earth is spinning around the North Pole axis such that if your right thumb is pointed in the direction of the north pole your fingers will cup in the direction the earth is rotating. If the point in question is located at 37.2° latitude, then solar noon occurs as the earth rotates around its axis so that the point in question is toward the sun. At that point in time, the angle between the sunlight and a normal to the surface of the earth will be 37.2°–23.44° = 13.76°. As the earth spins on its axis, regions near the South Pole will never see the sun. Regions near the North Pole will have 24-h daylight.

FIGURE 7.8 Winter Solstice.

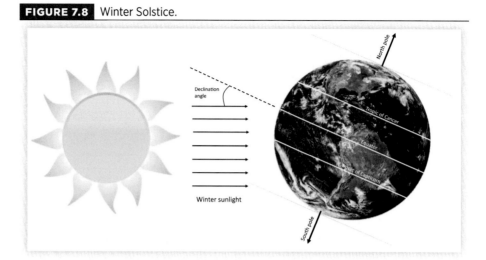

Figure 7.8 illustrates the winter solstice (December 21) when the earth has rotated around the sun, so it is now on the right side of the sun as illustrated in the figure. At the winter solstice it appears from earth that the sun is directly over the Tropic of Capricorn. If the point in question is located at 37.2° latitude, then solar noon occurs as the earth rotates around its axis so that the point in question is toward the sun. At that point in time the angle between the sunlight and a normal to the surface of the earth will be 37.2° + 23.44° = 60.64°. As the earth spins on its axis regions near the North Pole will never see the sun. Regions near the South Pole will have 24-h daylight.

Two times each year, September 21 and March 21, the earth is in equinox position. It is hard to draw a good diagram for the equinox position. Imagine that the earth is located above the page directly over the sun. It will appear from earth that the sun is directly over the equator during the equinox. Northern and southern hemispheres will get the same amount of daylight at the equinox.

7.5.2 **Hour Angle**

The declination angle can be used to find the sun angle at solar noon, but as the earth rotates on its axis the angle between the solar array and the sun varies with time of day. The hour angle accounts for the movement of sun during the day. The hour angle depends on the longitude location. The earth rotates 15° per hour, which makes it appear from earth that the sun moves 15° per hour around the earth. If we use Greenwich Mean Time (GMT) as the time, the longitudinal location of the sun is at −15* (GMT-12). The negative is because the sun moves in the negative longitude direction, and the −12 is because the longitudinal location of the sun is at Greenwich, England at noon GMT. Missouri is in the central time zone, which is 6 h behind GMT. Eastern time is 5 h behind GMT and Pacific time is 8 h behind GMT.

Example 7.5: Find the longitude location of the sun at 10:00 am (regular time) in Missouri.

    Solution: The sun is at −15 × (10 + 6−12) = −60° longitude. Please notice that if we are on daylight savings time the central time zone is only 5 h behind GMT and the longitude location of the sun would be −45° longitude.

Example 7.6: Springfield, Missouri, is at −93.29156° longitude. What is the longitudinal angle between the sun and Springfield at 10:00 am standard time?

    Solution: To solve this problem take the longitude position of Springfield and subtract the longitude position of the sun. At 10:00 am the longitude angle is −93.29156° - (−60°) = −33.29156°. The negative sign indicates that the sun will appear to be east of Springfield in the sky. The hour angle for Springfield, Missouri, at 10:00 am standard time is -33.29156°.

7.5.3 Hour Angle Definition

The hour angle is defined as the longitude position of the point on earth minus the longitude position of the sun (city longitude - sun longitude position). A negative hour angle means the sun will appear to be east of the city. An hour angle of zero is solar noon for the city. A positive hour angle means the sun will appear to be west of the city.

    The power that will be produced by a solar array is proportional to the cosine of the angle between the solar array and the sunlight. For the derivation we will assume that the solar array is parallel to the surface of the earth. Stationary arrays are usually tilted toward the south to capture more solar energy. Once the formula and process are developed for solar arrays parallel to the earth surface, it is relatively easy to incorporate the angle of tilt.

    In the derivation, the sun is located by the declination angle θ_D and the hour angle θ_H. The position of the solar array on the earth is located from its Latitude φ_{Lt} and Longitude φ_{Lo}. Define a coordinate system so that the z-axis is the North Pole axis. The x-axis comes from the center of the earth through the equator directly south of where the solar array is located. The hour angle is the east or west location of the sun relative to where the solar array is located. $\mathbf{n}_E$ is a unit normal vector perpendicular to the surface of the earth where the solar array is located. **Figure 7.9** illustrates the coordinate system used.

    With the coordinate system defined, the unit normal to the surface of the earth $\mathbf{n}_E$ is defined as:

$$\mathbf{n}_E = \cos\left(\varphi_{Lt}\right)\mathbf{i} + 0\,\mathbf{j} + \sin\left(\varphi_{Lt}\right)\mathbf{k} \tag{7.27}$$

The unit normal vector for the sun depends on the declination angle θ_D and the hour angle θ_H. The z-component of the normal vector of the sun is the sine of the declination angle. The sun normal is then projected into the x-y plane and the projection is divided into x and y components. The unit normal vector of the sun $\mathbf{n}_S$ is then defined as:

$$\mathbf{n}_S = \cos\left(\theta_D\right)\cos\left(\theta_H\right)\mathbf{i} + \cos\left(\theta_D\right)\sin\left(\theta_H\right)\mathbf{j} + \sin\left(\theta_D\right)\mathbf{k} \tag{7.28}$$

The dot product of $\mathbf{n}_E \cdot \mathbf{n}_S$ is the cosine of the angle between the sunlight and the normal to the surface of the earth. Since the power produced by the solar array is

FIGURE 7.9 Sun-Earth position and vectors.

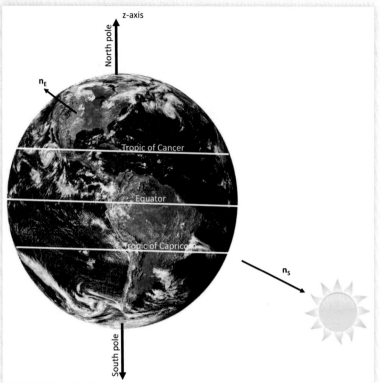

proportional to the cosine of the angle, the parameter needed is the cosine of the angle. Defining θ as the angle between the sunlight and the solar array it follows:

$$\cos(\theta) = \cos(\theta_D)\cos(\theta_H)\cos(\varphi_{Lt}) + \sin(\theta_D)\sin(\varphi_{Lt}) \tag{7.29}$$

Equation 7.29 is the factor that will be used to multiply the solar intensity for a solar array that is mounted parallel to the ground. That is, if the solar intensity is 950 W/m², the intensity of the sunlight striking the solar array will be (950 W/m²) cos(θ). The process is to use the day of the year (N) and Eq. 7.22 to find the declination angle θ_D. The time of day is used to find the longitudinal location of the sun. The hour angle θ_H is the longitudinal location of the city minus the longitudinal location of the sun. Equation 7.29 is then used to find cos (θ).

To model a solar car traveling along a road, we would approximate the solar array to be parallel to the ground. The intensity of the sunlight striking the solar array would be proportional to the cosine of the angle between a normal to the surface of the earth and the angle of the sun as shown in Eq. 7.29. The model developed is useful for modeling the solar array power received by a solar powered car.

The next step is to develop a spreadsheet model for the cosine factor for a particular location on earth (latitude and longitude) and the time of day. This model gives a good approximation of solar intensity during the day for a solar array that is parallel to the ground on a clear sunny day. The solar constant at sea level is ~950 W/m². The constant varies slightly with time of year and cycles of the sun. Solar intensity increases with altitude to a maximum of ~1400 W/m² above the atmosphere.

Example 7.7: Assume the declination angle (θ_D) was calculated to be 20° on the day in question. Assume the solar array is in Springfield, Missouri at latitude 37.2° and longitude -93.2916°. Develop a spreadsheet model for the solar factor ($\cos(\theta)$) of Eq. 7.29. Plot the solar factor as a function of time of day.

Solution: The first column in the spreadsheet is time of day, GMT, with a time of 0 representing noon GMT. The second column is the longitude location of the solar array. The third column is the sun location. The longitudinal location of the sun moves 15° per hour as shown in the spreadsheet. Column 4 is the hour angle θ_D, which is column 2 minus column 3. Columns 5 and 6 are the declination angle and latitude location of the solar array, which are constant for this problem. The Cosine Factor column is calculated using Eq. 7.29. Springfield is in the central time zone, which is 6-h behind the time in Greenwich. On standard time it would be 6:00 am in Springfield, or on daylight savings time it would be 7:00 am. The column for Springfield Time is included so the reader can relate the solar factor to local time. The Solar Factor column is equal to the Cosine Factor column except when the Cosine Factor is negative. A negative value in Eq. 7.29 indicates the sun has set. A logic statement was added for the Solar Factor column to set it equal to zero when the Cosine Factor is negative. **Table 7.4** is the spreadsheet.

TABLE 7.4 Spreadsheet model for solar intensity.

GMT with noon=0	Longitude	Sun Angle	Hour Angle	Declination Angle	Latitude	Cosine Factor	Springfield Time	Solar Factor
0	-93.2916	0	-93.2916	20	37.2	0.163808	6	0.163808
1	-93.2916	-15	-78.2916	20	37.2	0.358677	7	0.358677
2	-93.2916	-30	-63.2916	20	37.2	0.543195	8	0.543195
3	-93.2916	-45	-48.2916	20	37.2	0.704787	9	0.704787
4	-93.2916	-60	-33.2916	20	37.2	0.832442	10	0.832442
5	-93.2916	-75	-18.2916	20	37.2	0.917458	11	0.917458
6	-93.2916	-90	-3.2916	20	37.2	0.954044	12	0.954044
7	-93.2916	-105	11.7084	20	37.2	0.939705	1	0.939705
8	-93.2916	-120	26.7084	20	37.2	0.875418	2	0.875418
9	-93.2916	-135	41.7084	20	37.2	0.765566	3	0.765566
10	-93.2916	-150	56.7084	20	37.2	0.617633	4	0.617633
11	-93.2916	-165	71.7084	20	37.2	0.441702	5	0.441702
12	-93.2916	-180	86.7084	20	37.2	0.249762	6	0.249762
13	-93.2916	-195	101.7084	20	37.2	0.054893	7	0.054893
14	-93.2916	-210	116.7084	20	37.2	-0.12963	8	0
15	-93.2916	-225	131.7084	20	37.2	-0.29122	9	0
16	-93.2916	-240	146.7084	20	37.2	-0.41887	10	0
17	-93.2916	-255	161.7084	20	37.2	-0.50389	11	0
18	-93.2916	-270	176.7084	20	37.2	-0.54047	12	0
19	-93.2916	-285	191.7084	20	37.2	-0.52613	1	0
20	-93.2916	-300	206.7084	20	37.2	-0.46185	2	0
21	-93.2916	-315	221.7084	20	37.2	-0.352	3	0
22	-93.2916	-330	236.7084	20	37.2	-0.20406	4	0
23	-93.2916	-345	251.7084	20	37.2	-0.02813	5	0
24	-93.2916	-360	266.7084	20	37.2	0.163808	6	0.163808

FIGURE 7.10   Plot of Solar Factor for Springfield, Missouri, May 20.

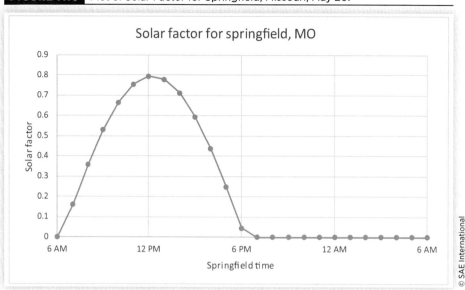

A declination angle of 20° corresponds to May 20 or July 24, plus or minus a day depending on leap year. Once the spreadsheet is developed it can be modified to work for any city where the solar array is located and for any day of the year. A plot of the solar factor is shown in Fig 7.10.

Figure 7.10 illustrates how the power produced by a solar array parallel to the ground would vary with time of day. The maximum power output would occur approximately at GMT 6:00 PM, which will be noon standard time in Springfield, or 1:00 PM daylight savings time. The output is lower at other times of the day. The spreadsheet model developed in Table 7.4 is a good starting point for modeling solar cars. The latitude, longitude, and time of day would vary as the car drives the race route. But with some thought and effort the concepts in Table 7.4 could be used to develop a model for solar cars driving along a race route.

7.5.4 Process for Stationary Tilted Arrays

Stationary solar arrays are usually tilted to the south to allow them to collect more solar energy. This changes the problem because the normal to the earth is no longer normal to the solar array. If the array is tilted to the south an amount θ_T, the normal to the solar array $\mathbf{n}_A$ is given by the following:

$$\mathbf{n}_A = \cos\left(\varphi_{Lt} - \theta_T\right)\mathbf{i} + 0\,\mathbf{j} + \sin\left(\varphi_{Lt} - \theta_T\right)\mathbf{k} \tag{7.30}$$

The angle for the sun $\mathbf{n}_S$ does not change from what it was for the array parallel to the ground. Taking the dot product $\mathbf{n}_A \cdot \mathbf{n}_S$ is the cosine factor for a tilted solar array. Unfortunately, it is not quite this simple. The simple dot product approach assumes the solar array is high enough above the surface of the earth that the earth is not blocking the sunlight from reaching the solar array. If the sun can reach the solar array the cosine factor can be obtained from the formula below.

$$\cos(\theta) = \cos(\theta_D)\cos(\theta_H)\cos(\varphi_{Lt} - \theta_T) + \sin(\theta_D)\sin(\varphi_{Lt} - \theta_T) \qquad (7.31)$$

This can be incorporated into the solar array spreadsheet that was developed for the array parallel to the ground. First obtain the solar factor for when the solar array is parallel to the ground as was done in the previous spreadsheet. This spreadsheet allows us to determine when the sun is above the horizon and able to shine on the solar array. A logic statement is added so that Eq. 7.31 is used to find the tilted array solar factor when the flat solar factor is not zero. It's a little confusing, but this is the best way to model tilted solar arrays.

For summer months in the northern hemisphere the sun rises in the northeast and sets in the northwest. In early morning and late afternoon, the tilted solar array produces less power than the flat solar array. In summer months there is not much difference overall in how much energy is collected during the day whether the solar array is flat or tilted to the south. In spring, fall, and winter, the tilted solar array will collect significantly more solar energy than a flat solar array. Mechanical systems have been used to adjust the tilt of the solar array during the day, and/or track the sun and keep the array angled at the sun during the day. These mechanical systems allow the solar array to collect more energy, but the mechanical complexity makes the system more expensive, less reliable, and prone to being damaged by the weather. Most solar panels are mounted in a fixed position and tilted to the south at an angle to maximize the amount of electricity they produce during the year. The ideal angle of tilt depends on the latitude location of the solar array. In home construction the most common roof used is the 4:12 pitch, which is an angle of 18.4°. If the solar panels are mounted directly on the south facing roof, they will be tilted 18.4° to the south, which is a good angle of tilt for most places in the USA. It is not the ideal angle of tilt, but when you consider the cost of mounting the cells on the roof and the fact that they are less likely to be damaged by the weather if mounted flat on the roof, this is a good design.

Eq. 7.31 was developed assuming the city is in the northern hemisphere so that φ_{Lt} is a positive latitude. Tilting the solar array to the south makes θ_T a positive angle. The normal situation in the northern hemisphere makes φ_{Lt} and θ_T positive angles. Equation 7.31 works in the southern hemisphere where φ_{Lt} is a negative latitude. In the southern hemisphere the solar array would be tilted to the north, making θ_T a negative angle. The normal situation in the southern hemisphere makes φ_{Lt} and θ_T negative angles, but Eq. 7.31 still yields the correct answer if the negative values are used.

A column can be added to the spreadsheet model in Table 7.4 titled "Tilted Solar Factor" to total the solar factors for the day. The tilt angle can be adjusted to maximize the total of the solar factors and the reader can discover the optimum tilt of the array to maximize the energy collected during the day. For a declination angle of 20° and a latitude of 37.2° it turns out that the optimum tilt is zero degrees. That is, the maximum energy would be collected by putting the solar array parallel to the ground. Tilting the solar array will increase the peak power in the middle of the day but will decrease the power in the morning and evening and overall the maximum energy will be collected with the array parallel to the ground.

TABLE 7.5 Spreadsheet model for tilted solar array.

Declination	0	degrees		Latitude	37.2	degrees		
				Tilt Angle	37.2	degrees		
GMT with noon = 0	Longitude	Sun Angle	Hour Angle	Declination Angle	Latitude	Cosine Factor	Flat Solar Factor	Tilt Solar Factor
0	−93.2916	0	−93.2916	0	37.2	0.163808	0	0
1	−93.2916	−15	−78.2916	0	37.2	0.358677	0.16164	0.202931
2	−93.2916	−30	−63.2916	0	37.2	0.543195	0.358	0.44945
3	−93.2916	−45	−48.2916	0	37.2	0.704787	0.529963	0.66534
4	−93.2916	−60	−33.2916	0	37.2	0.832442	0.66581	0.835888
5	−93.2916	−75	−18.2916	0	37.2	0.917458	0.756282	0.949472
6	−93.2916	−90	−3.2916	0	37.2	0.954044	0.795216	0.99835
7	−93.2916	−105	11.7084	0	37.2	0.939705	0.779957	0.979193
8	−93.2916	−120	26.7084	0	37.2	0.875418	0.711545	0.893306
9	−93.2916	−135	41.7084	0	37.2	0.765566	0.594642	0.746541
10	−93.2916	−150	56.7084	0	37.2	0.617633	0.437215	0.5489
11	−93.2916	−165	71.7084	0	37.2	0.441702	0.249994	0.313853
12	−93.2916	−180	86.7084	0	37.2	0.249762	0.045735	0.057418
13	−93.2916	−195	101.7084	0	37.2	0.054893	0	0
14	−93.2916	−210	116.7084	0	37.2	−0.12963	0	0
15	−93.2916	−225	131.7084	0	37.2	−0.29122	0	0
16	−93.2916	−240	146.7084	0	37.2	−0.41887	0	0
17	−93.2916	−255	161.7084	0	37.2	−0.50389	0	0
18	−93.2916	−270	176.7084	0	37.2	−0.54047	0	0
19	−93.2916	−285	191.7084	0	37.2	−0.52613	0	0
20	−93.2916	−300	206.7084	0	37.2	−0.46185	0	0
21	−93.2916	−315	221.7084	0	37.2	−0.352	0	0
22	−93.2916	−330	236.7084	0	37.2	−0.20406	0	0
23	−93.2916	−345	251.7084	0	37.2	−0.02813	0	0
24	−93.2916	−360	266.7084	0	37.2	0.163808	0	0
						Total	6.085999	7.640641

If the declination angle is zero degrees (March 21 and September 21) and the latitude is 37.2° then the solar array will collect more energy if it is tilted to the south. The optimum tilt for this case is 37.2°, which is the latitude location. If the declination angle is zero the ideal tilt for the solar array is equal to the latitude location of the solar array. The spreadsheet for this case is shown in **Table 7.5**. Tilting the array to 37.2° makes the total solar factor for the day 7.64, whereas keeping the array parallel to the ground give a total solar factor of 6.89. Tilting the solar array will increase the energy recovered by 25.5%.

If the declination angle is near −20° (December and early January), the maximum energy for the day will be collected by tilting the solar array 63.5°. If the declination angle is near +20° (June and early July) it is best to have the array parallel to the ground. To summarize for a latitude of 37.2°, the array should be parallel to the ground in summer, tilted to the south 37.2° in spring and fall when the sun is at the equinox and tilted 63.5 degrees to the south in the winter. This model is developed assuming the solar array is tilted to the south. Tilting the array in other directions or having a mechanism to allow the array to track the sun will yield different results.

7.6 **Solar Array Power Equation**

The model developed in this section is based on research done by Courtney Green in 2006, supervised by Doug Carroll [5]. There is extensive solar data available from the National Renewable Energy Lab (NREL). The data were collected at five different locations around the country (Elizabeth City State University in North Carolina, Central Oklahoma, Bluefield State College in Virginia, and two locations in Colorado). A model was developed with parameters, and the parameters were adjusted to fit the experimental data. The best-fit parameters are shown below the equation. The model gives the solar intensity of the sunlight P if the normal to the solar array is aimed directly at the sun at mid-day.

$$P = \left[P_\infty + P_V \cos\left(\frac{360(N-3)}{365} \right) \right] \left(\frac{P_0}{P_\infty} \right)^{\frac{t-h}{t(\cos(\theta))^m}} \tag{7.32}$$

In using Eq. 7.32 the input parameters are N = the day of the year, where January 1 is 1, and h = height above sea level in meters. The declination angle θ_D is calculated using Eq. 7.22 and the hour angle θ_H is calculated using the method in Section 7.5. The other parameters in the model were adjusted to achieve the best fit to the data available from NREL. The best fit parameters are listed below:

- P_∞ = 1400 W/m^2
- P_0 = 1087.445 W/m^2
- P_V = 116.4844 W/m^2
- t = 10304.022 m
- m = 0.65
- $\cos(\theta) = \cos(\theta_D)\cos(\theta_H)\cos(\varphi_{Lt}) + \sin(\theta_D)\sin(\varphi_{Lt})$

In developing the model, the parameters had a physical meaning, but their values were adjusted to make the model fit the data. P_∞ represents the solar intensity in space above the atmosphere. P_0 represents the solar intensity at sea level. P_V represents the variability of solar intensity with the time of year since the earth is closest to the sun in January and furthest from the sun in July. The parameter t represents what the thickness of the atmosphere would be if the gasses in the atmosphere were compressed to sea level density. The parameter m is empirical and helped us get a better fit to the data. Using these parameters gives a good overall fit between the model and the available data.

To get the power from the solar array, multiply the power from Eq. 7.32 by the cosine of the angle between the solar array and the sunlight. For solar arrays parallel to the ground use Eq. 7.29 and for solar arrays tilted to the south use Eq. 7.31.

Example 7.8: Assume a sunny day April 27, 2017 and calculate the power output for a 20 square meter solar array tilted 18.4 degrees to the south. The solar array is 18% efficient in converting the sunlight into electricity. Assume latitude ϕ_{Lt} = 30.32° and the hour angle θ_H = 23.5° for the time in question. Assume the solar array is located at an altitude of 1400 ft = 426.7 m above sea level.

Solution: Equation 7.32 is complex, and I prefer to evaluate it a piece at a time and then put the pieces together. First find the day of the year for April 27, 2017 and the first part of the equation:

$$N = 31 + 28 + 31 + 27 = 117 \tag{7.33}$$

$$P_\infty + P_V \cos\left(\frac{360(N-3)}{365}\right) = 1355.5 \text{ W/m}^2 \tag{7.34}$$

When taking the cosine in Eq. 7.34 the angle is in degrees (not radians). Equation 7.22 is used to calculate the declination angle. θ_D = 13.8296° on April 27, 2017. The latitude is 30.32° and the hour angle is 23.5°. The cos (θ) term in Eq. 7.32 is calculated as:

$$\cos(\theta) = \cos(13.8296)\cos(23.5)\cos(30.32) + \sin(13.8296)\sin(30.32) = 0.889347 \tag{7.35}$$

The altitude correction in Eq. 7.32 is computed as:

$$\left(\frac{P_0}{P_\infty}\right)^{t\left((\cos(\theta))^m\right)^{\frac{t-h}{t}}} = \left(\frac{1087.445}{1400}\right)^{\frac{10304.022-426.7}{10304.022\left(0.889347^{0.65}\right)}} = 0.770003 \tag{7.36}$$

Equations 7.34 and 7.36 are combined to find the solar power striking a solar array with the normal to the solar array pointed directly at the sun as:

$$P = \left(1355.5 \text{ W/m}^2\right)(0.770003) = 1044 \text{ W/m}^2 \tag{7.37}$$

The model predicts that the solar intensity at latitude of 30.32° and altitude of 1400 ft above sea level on April 27, 2017, the solar intensity striking a solar array aimed directly at the sun will be 1044 W/m². The solar array in question is tilted 18.4° to the south, so it is not aligned with the sun. Equation 7.31 is used to account for the misalignment with the sun:

$$\cos(\theta) = \cos(13.8296)\cos(23.5)\cos(30.32-18.4)$$
$$+ \sin(13.8296)\sin(30.32-18.4) = 0.920646 \tag{7.38}$$

Remember that the tilted solar array factor can only be positive if the flat solar array factor calculated in this case in Eq. 7.35 is positive. The logic statement discussed before is still required for the tilted solar array equation. In this case it is positive, so the 0.920646 factor calculated in Eq. 7.38 is valid. The solar array has an area of 20 m² and the solar cells are 18% efficient in converting the sunlight into electricity. The solar array power is computed as:

$$\text{Array Power} = \left(1044 \text{ W / m}^2\right)(0.920646)\left(20 \text{ m}^2\right)(0.18) = 3460 \text{ W} \qquad (7.39)$$

This is the solar array power when the hour angle is 23.5°, which would be ~1:30 pm standard time. The exact time depends on where the solar array is in the time zone.

References

1. https://www.e-education.psu.edu/ebf200/node/151

2. https://www.eia.gov/state/seds/data.php?incfile=/state/seds/sep_prices/total/pr_tot_US.html&sid=US

3. https://en.wikipedia.org/wiki/Table_Rock_Lake

4. https://en.wikipedia.org/wiki/Position_of_the_Sun

5. Green, Courtney, "Solar Power Model," in *Proceedings of Opportunities for Undergraduate Research*, University of Missouri-Rolla, 2006.

CHAPTER 8

Greenhouse Gasses and How They Impact the Temperature of the Earth

8.1 Radiative Heat Transfer of Earth

Climatologists have been saying that greenhouse gasses (GHG) will increase the temperature of the earth and lead to global warming. There is concern that the carbon dioxide emitted when burning fossil fuel will lead to global warming. The first step to understand how GHG impacts the temperature of the earth is to look at the radiative heat transfer of the earth as a whole.

The earth receives heat from the sun as electromagnetic radiation. Most of the energy from the sun is in the visible range because of the temperature of the sun. Objects emit radiation in a range of energy that is related to their temperature with hotter objects emitting higher energy radiation and cooler objects emitting lower energy radiation. The temperature of the sun causes it to emit radiation mostly in the visible range, with smaller amounts of radiation in higher and lower energy ranges.

The flame from a campfire is very hot and emits radiation in the visible range. This allows us to see the flame. Once the fire burns down there are still glowing red and orange coals that are hot enough to emit radiation in the lower energy frequencies of the visible range. Within the visible range, red and orange are lower energy frequencies, and blue and violet are higher energy frequencies. Once the fire burns down completely it does not emit visible light anymore, but it is still hot. The fire has cooled to the point that it is emitting radiation in the infrared range. We cannot see infrared radiation, but we can feel the heat.

The surface of the earth emits radiation in the infrared range. The earth receives heat energy from the sun mostly in the visible range of frequencies and emits energy back into space as infrared energy. For energy balance the earth must receive the same amount of energy from the sun that it emits as infrared radiation. If the earth was to receive more energy from the sun than it emits the earth would become warmer. Conversely if the earth were to receive less energy from the sun than it emits it would become cooler. We see this happen every day. The earth will tend to warm during the day when it receives lots of solar energy and will tend to cool at night. The average solar radiation [1] striking the top of the atmosphere is 1361 W/m². Solar radiation is not constant because the radiation emitted by the sun is not constant. The earth is closer to the sun in January and further from the sun in July, which also impacts the intensity of the solar radiation. But if we average through the year the solar intensity is 1362 W/m².

Sunlight strikes the earth in profile so that if R is the radius of the earth (plus atmosphere) the total energy striking earth is (1362 W/m²) πR^2. The earth will emit infrared radiation in all directions from the surface of the earth, and the surface of the earth (plus atmosphere) is a sphere of surface area $4 \pi R^2$. To convert the 1362 W/m² striking a circular profile area to an average radiation over the spherical surface of the earth, we need to ratio the areas. The average solar radiation striking the top of the atmosphere is:

$$\text{Average Solar Radiation} = \left(1361 \text{ W/m}^2\right)\left(\frac{\pi R^2}{4 \pi R^2}\right) = 340.5 \text{ W/m}^2 \qquad (8.1)$$

If 100% of the solar energy striking the top of the atmosphere were absorbed by the earth and atmosphere, then the earth would receive 340.5 W/m² average solar radiation over its surface all the time. Some parts of the earth receive more solar radiation than others and this causes some parts of the earth to be warmer than others. Assuming 100% absorption of the energy from the sun, the average solar radiation would be 340.5 W/m². But not all radiation from the sun is absorbed by the earth.

Some of the energy from the sun is reflected into space by the clouds and the surface of the earth. The reason we can see the earth from space is because of the reflected sunlight. If all sunlight was absorbed by the earth and atmosphere it would not be possible to see the earth from space. To do the radiative-heat transfer we need to know how much solar energy is absorbed by the earth and atmosphere and how much is reflected into space. Through experimental work, atmospheric scientists have found that on average the earth absorbs 239 W/m² of radiation [2].

$$\text{Average Solar Radiation Absorbed} = 239 \text{ W/m}^2 \left(\text{Experimental Value}\right) \qquad (8.2)$$

This means that 340.5–239 = 101.5 W/m² is being reflected into space. In percentages, 70.2% of the sunlight is absorbed by the earth and atmosphere and 29.8% is reflected into space. GHG do not absorb sunlight, so increasing the carbon dioxide or other GHG in the atmosphere will not change the amount of solar radiation that is absorbed. The average solar radiation earth receives from the sun will not change as the amount of GHG increase in the atmosphere.

The next step in understanding the radiative heat transfer of the earth is to understand how the earth emits infrared radiation back into space. The average temperature of the earth is constant. Scientists have averaged temperature readings over the surface of the earth and measured the average earth temperature as 288 K, which is 59 °F.

That is, if you average the temperature over the entire earth, day and night and throughout the year, the average temperature is 288 K.

The Stefan-Boltzmann relationship says that a body emits radiation according to its temperature raised to the fourth power. The Stefan-Boltzmann constant in its metric form is 5.67×10^{-8} W/m^2 °K^4. As the average earth temperature is 288 K, the infrared radiation from earth is calculated as:

$$\text{Earth Surface Radiation} = \left(\frac{5.67 \times 10^{-8} \text{ W}}{\text{m}^2 \text{°K}^4} \right) \left(288\text{°K} \right)^4 = 390 \text{ W/m}^2 \qquad (8.3)$$

Some of the infrared radiation from earth is absorbed by GHG. For the average earth temperature to be constant, the energy absorbed by the earth and atmosphere from the sun (239 W/m^2) must equal the infrared energy leaving the earth from the top of the atmosphere. The earth's surface is emitting 390 W/m^2 infrared energy, so the difference must be absorbed by the GHG in the atmosphere. The energy balance yields:

$$\text{GHG Absorption} = 390 \text{ W/m}^2 - 239 \text{ W/m}^2 = 151 \text{ W/m}^2 \qquad (8.4)$$

In percentages, 61.3% of the infrared radiation from the surface of the earth will find its way through the atmosphere and into space. In total 38.7% of the infrared radiation will be absorbed by the GHG in the atmosphere. **Figure 8.1** is helpful in illustrating the energy balance of the earth.

The most common GHG are water vapor, carbon dioxide, methane, halocarbons (CFCs), NOx, and ozone. Water vapor is the strongest GHG as far as absorbing infrared radiation. The amount of water vapor in the atmosphere depends on the average temperature of the atmosphere. Warmer air will hold more moisture than cooler air. The moisture in the air is somewhat self-regulating because an excess of moisture in the air will cause rain, which removes moisture from the atmosphere. Carbon dioxide is the second most

FIGURE 8.1   Radiative energy balance for earth.

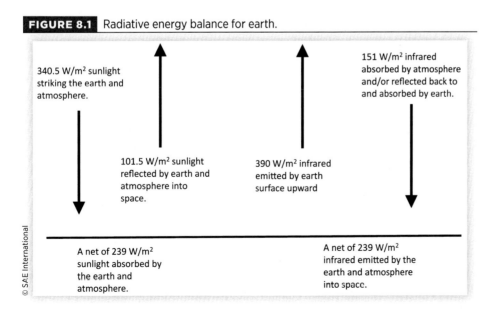

340.5 W/m² sunlight striking the earth and atmosphere.

151 W/m² infrared absorbed by atmosphere and/or reflected back to and absorbed by earth.

101.5 W/m² sunlight reflected by earth and atmosphere into space.

390 W/m² infrared emitted by earth surface upward

A net of 239 W/m² sunlight absorbed by the earth and atmosphere.

A net of 239 W/m² infrared emitted by the earth and atmosphere into space.

© SAE International

important GHG, and we are concerned about the increase of carbon dioxide in the atmosphere causing the earth to warm. Methane is more effective at absorbing infrared radiation than carbon dioxide, but the concentration of methane in the atmosphere is very small, so overall methane absorbs less infrared radiation than carbon dioxide. Some scientists are concerned that thawing of the polar regions will release massive quantities of methane into the atmosphere, which would greatly increase the amount of energy absorbed by GHG. Halocarbons, NOx, and ozone all contribute to GHG absorption, but to a lesser extent than water vapor, carbon dioxide, and methane.

Example 8.1: Assume that an increase in GHG in the atmosphere allows the atmosphere to absorb 40% of the infrared radiation from earth rather than the 38.7% it is absorbing now. How will this affect the average temperature of the earth?

Solution: As the temperature of the earth increases, the earth will emit more infrared radiation due to the temperature increase. If GHG absorb 40% of the infrared radiation, then 60% will leave the atmosphere and go into space. The earth will still absorb 239 W/m^2 from the sun, and in order for the earth to stabilize to a new temperature the radiation leaving the top of the atmosphere going into space must equal the 239 W/m^2. The radiation from the surface of the earth is calculated as:

$$\text{Earth Surface Radiation} = \frac{239 \text{ W/m}^2}{0.60} = 398.3 \text{ W/m}^2 \tag{8.5}$$

The Stephan-Boltzmann relationship can then be used to determine the average earth temperature required to emit 398.3 W/m^2 radiation from the surface of the earth.

$$\left(\frac{5.67 \times 10^{-8} \text{ W}}{\text{m}^2 {}^\circ\text{K}^4} \right) (T)^4 = 398.3 \text{ W/m}^2 \tag{8.6}$$

Solving Eq. 8.6 the earth's average temperature is 289.5°K. The solution shows that increasing the absorptivity of the atmosphere from 38.7% to 40% will increase the average temperature of the earth by 1.5°K (or 1.5°C). Another way of looking at it is to say increasing the absorption of the atmosphere from 151 W/m^2 to 156 W/m^2 will cause the 1.5°C temperature increase. An average temperature increase of 1.5°C will significantly change our environment. The purpose of developing this model was to show that a relatively small increase in the absorptivity of the atmosphere will cause significant climate change.

The radiation model presented in this chapter is a great oversimplification. In reality, the temperature on earth varies significantly from place to place. The model uses the average earth temperature and assumes the temperature is constant over the surface of the earth. The earth receives sunlight only during daytime hours and there is a significant variation in solar intensity on the surface of the earth. The model uses the average solar radiation and assumes the radiation is spread evenly over the surface of the earth. It is easy to dismiss the model developed here as a great oversimplification of what is actually happening. The advantage of the model developed in this chapter is that it is relatively simple, and most scientists and engineers can do the calculations and make a good estimate of how much GHG will impact the average earth temperature. Climate scientists have worked through much more detailed models than what has been presented here. The results of this simple model are similar to the results of the more complex models. Though not perfect, the simple model developed in this unit can be used to estimate how GHG affects the average temperature of the earth.

8.2 Carbon Dioxide Portion of GHG Absorption

Carbon dioxide is the second most important GHG in the atmosphere. Water vapor accounts for more of the 151 W/m² infrared radiation absorbed by the atmosphere than all the other GHG combined. It is difficult to get an accurate measurement of the amount of energy absorbed by individual GHG because they are mixed in the atmosphere. We can do experiments with individual GHG and measure the amount of infrared radiation that will be absorbed, but we cannot simply add the contributions of the different GHG. An infrared photon absorbed by water vapor will not also be absorbed by carbon dioxide. It is absorbed by one or the other but not both. If we test the individual GHG and add the results, we are double counting some of the absorption and it leads to a high estimate of the infrared energy absorbed.

Recognizing the limitations discussed in the paragraph above, it is possible to estimate the amount of energy absorbed by carbon dioxide if it were the only GHG in the atmosphere. This model yields a high estimate of the energy absorbed by carbon dioxide, but it is a place to start the discussion. Stull et al. measured the infrared energy transmitted through carbon dioxide as a function of the carbon dioxide concentration [3]. The concentration of carbon dioxide was measured in mole/m², and the device measured the fraction of infrared radiation transmitted through the experimental chamber. The results are presented in **Table 8.1**.

We are more interested in the fraction of energy absorbed, which is one minus the fraction transmitted. The absorption increases as the carbon dioxide concentration increases. The data seem to follow a logarithmic function. A good fit to the data is:

$$\text{Absorption Fraction} = (0.09083)\ln(\text{Concentration}) + 0.0681 \qquad (8.7)$$

TABLE 8.1 Results from Stull et al.

Concentration of carbon dioxide (mole/m²)	Fraction of energy transmitted through
40.9	0.5917
81.8	0.5328
204.5	0.4535
409	0.3810

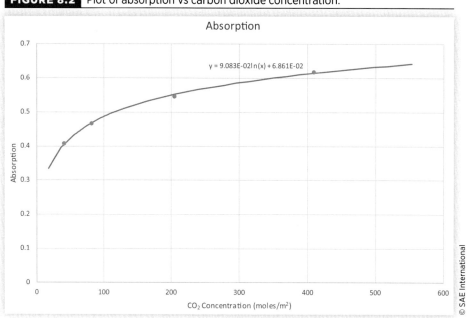

FIGURE 8.2 Plot of absorption vs carbon dioxide concentration.

The logarithm in Eq. 8.7 is the natural log (not the base 10 log). **Figure 8.2** illustrates how the absorption varies with the concentration of carbon dioxide. The orange line is Eq. 8.7 and the blue dots are the data points.

Carbon dioxide can absorb a range of frequencies of infrared radiation with wavelengths in the 13-17 μm range. Approximately 48.9% of the infrared radiation from earth falls in this range. When we look at the amount of infrared radiation absorbed, we need to multiply the equation by 0.489 because most of the infrared radiation emitted from the surface of the earth cannot be absorbed by the carbon dioxide.

The carbon dioxide molecules will absorb the infrared photons and then re-emit photons in a random direction. Approximately half of the re-emitted photons will travel up and escape the atmosphere and approximately half will travel downward and be absorbed. Equations 8.4-8.9 must also be multiplied by a factor of ½ to account for this.

The next step is to calculate the concentration of carbon dioxide on earth in moles per square meter surface area of the earth. There are 1.78×10^{20} moles of gas in the atmosphere. The radius of the earth is R = 6,370,000 m. Dividing the moles of gas in the atmosphere by the surface area of the earth yields:

$$\frac{1.78 \times 10^{20} \, \text{mole}}{4\,\pi \left(6,370,000 \, \text{m}\right)^2} = 349,085 \, \text{mole/m}^2 \tag{8.8}$$

This is the total number of gas molecules per square meter of surface area on the earth. We keep track of the carbon dioxide in parts-per-million (ppm). The atmosphere currently has a little over 400 ppm carbon dioxide, which means that of every million molecules in the atmosphere, a little over 400 are carbon dioxide molecules. If we define

C_{ppm} as the parts per million carbon dioxide in the atmosphere, then the CO_2 concentration in moles per square meter earth surface is calculated as:

$$CO_2 \text{ Concentration} = (0.349)C_{ppm} \text{ mole/m}^2 \qquad (8.9)$$

Putting all of this together, the energy absorbed is calculated as:

$$\text{Energy Absorbed by } CO_2 = \frac{1}{2}(0.489)(390 \text{ W/m}^2)\left[(0.09083)\log\left((0.349)C_{ppm}\right)+0.0681\right]$$

$$(8.10)$$

As an example calculation, if there are 400 ppm CO_2 in the atmosphere, the CO_2 concentration is calculated from Eq. 8.9 as (0.349) (400) = 139.2 mole/m². This is within the range of experimental values measured by Stull et al., so it is reasonable to use their results. What this means is the infrared radiation from 1 m² of surface area on the earth will travel upward through the atmosphere where there are 139.2 moles CO_2 that might absorb the infrared radiation. The earth at 288 K emits 390 W/m² infrared radiation. At 400 ppm carbon dioxide the energy absorbed by the carbon dioxide in the atmosphere can be calculated using Eq. 8.10 as 49.3 W/m².

According to the model developed, the carbon dioxide will absorb 49.3 W/m² of the total 151 W/m² absorbed by the atmosphere. As a percentage, the carbon dioxide absorbs 32.6% of the energy that is absorbed by the atmosphere. As stated above, the model yields a high estimate because some of the infrared radiation that could be absorbed by carbon dioxide will be absorbed by water vapor and other GHG. The carbon dioxide will actually absorb less than the 49.3 W/m² calculated.

Water and other GHG absorb infrared radiation in different bands of frequencies or wavelengths. There is overlap in the frequency bands so that more than one GHG can absorb some frequencies of infrared radiation. These things make it difficult to determine exactly how much energy is absorbed by the carbon dioxide in the atmosphere, and how much is absorbed by other GHG.

As the CO_2 in the atmosphere is near 400 ppm, it makes sense to take the derivative of Eq. 8.10 and evaluate it at 400 ppm. This gives the slope of the curve at 400 ppm and allows us to estimate how small changes in C_{ppm} impact the average temperature of the earth. The derivative of Eq. 8.10 evaluated at C_{ppm} = 400 ppm is 0.0217. This means that increasing the CO_2 in the atmosphere by 1 ppm will increase the infrared radiation absorbed by 0.0217 W/m². This information can be used to estimate how increasing the carbon dioxide in the atmosphere will impact global warming.

Example 8.2: Assume that at 400 ppm CO_2 in the atmosphere the earth is at an average temperature of 288 °K, the surface of the earth is emitting 390 W/m², and that 151 W/m² is being absorbed by the atmosphere. Assume that increasing the CO_2 in the atmosphere 1 ppm will increase the energy absorbed by the atmosphere by 0.0217 W/m². Estimate the amount of global warming caused by increasing the CO_2 to (a) 450 ppm and (b) 500 ppm.

Solution: An increase to 450 ppm will increase the energy absorbed by (50) (0.0217) = 1.085 W/m². The atmosphere will absorb 151 + 1.085 = 152.085 W/m². To balance the energy the earth surface must radiate 390 + 1.085 = 391.085 W/m². The Stephan-Boltzman equation is then used to calculate the temperature of the surface of the earth:

$$\left(\frac{5.67 \times 10^{-8}\ \text{W}}{\text{m}^2 \text{°K}^4}\right)(T)^4 = 391.085\ \text{W/m}^2 \tag{8.11}$$

Solving Eq. 8.11 yields T = 288.185°K. From this model increasing the CO_2 in the atmosphere to 450 ppm would cause a 0.185°C increase in temperature. Increasing to 500 ppm would cause the atmosphere to absorb 151 + 2.17 = 153.17 W/m². The increase in global temperature would be 0.385°C.

8.3 Keeling Curve

Dr. Ralph Keeling began taking measurements of carbon dioxide in the atmosphere in 1958 at the Mana Loa observatory in Hawaii [4]. The data are plotted in **Figure 8.3**.

The average level of CO_2 in the atmosphere has increased each year. The fluctuations during the year are because there is more land area and more plants in the northern hemisphere than in the southern hemisphere. During summer in the northern hemisphere the plants will remove carbon dioxide from the atmosphere. As plants die in the winter the carbon dioxide in the atmosphere increases. This cycle and the upward trend have been going on since we began taking measurements in 1958.

If you take data close to the earth surface where there are many plants around, the plants will have a large impact on carbon dioxide concentration during the day. That is, there will be significant daily fluctuations. Mana Loa was chosen because there are few plants around the site, and it is high enough to measure a more consistent concentration

FIGURE 8.3 Keeling Curve.

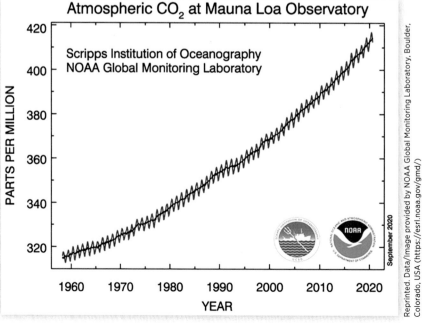

in the atmosphere. There are many other sites on earth where carbon dioxide concentration is measured now, and all sites are showing the concentration in the atmosphere is increasing.

The slope of the Keeling curve increased significantly in the mid-1990s. During the 1960s the carbon dioxide was increasing at a rate of about 1 ppm each year. In the most recent 20 years the carbon dioxide has been increasing at a rate of a little over 2 ppm each year. Worldwide we burn enough fossil fuel annually to amount to 4.5 ppm in the atmosphere. Approximately half of the carbon dioxide generated from burning fossil fuels goes into the atmosphere and about half is absorbed by the oceans.

Some projections assume the carbon dioxide concentration will continue to increase at 2 ppm, or that it accelerates its increase as we continue to burn an increasing amount of fossil fuel. It's hard to predict the future. Plants remove a significant amount of carbon dioxide in the atmosphere. The Keeling curve shows significant seasonal variation in carbon dioxide content because most of the land masses are in the northern hemisphere. Carbon dioxide content decreases when it is summer in the northern hemisphere because plants are removing lots of carbon dioxide from the atmosphere. Carbon dioxide increases when it is winter in the northern hemisphere because there are fewer plants on earth removing carbon dioxide from the atmosphere. A warmer earth will allow more plants to grow in Canada and Siberia, which are huge land masses. A large increase of plant life on earth will reduce the amount of carbon dioxide in the atmosphere. There may come a point where the plants take more carbon dioxide out of the atmosphere than humans put into the atmosphere. On the other side of the argument people say that other parts of the world will become so hot plants will die, which would tend to increase carbon dioxide content in the atmosphere. The Keeling curve has shown a steady increase in carbon dioxide, but that steady increase may or may not continue. In the near term it is likely that the increase in carbon dioxide will continue but projecting out 100 years is a projection of unknown accuracy.

The Keeling curve can be combined with the models above to estimate the increase in earth temperature due to rising levels of carbon dioxide in the atmosphere. In the year 2000 there were 368 ppm carbon dioxide and the level has been increasing at 2.2 ppm annually since then. If we extrapolate this into the future the carbon dioxide in the atmosphere (Cppm) is:

$$C_{ppm} = (year - 2000)(2.2) + 368 \qquad (8.12)$$

Once this is known, Eq. 8.10 is modified to include that the emissivity of the earth increases as the atmosphere absorbs more infrared radiation. That is, as the atmosphere absorbs more infrared radiation, the average infrared radiation from the surface of the earth will increase to values above 390 W/m² to balance the energy emitted by and received by the earth. Letting "Emissivity" be the average emissivity of the surface of the earth, Eq. 8.13 is the modified version of the equation.

$$\text{Energy Absorbed by } CO_2 = \frac{1}{2}(0.489)(\text{Emissivity})\left[(0.09083)\log\left((0.349)C_{ppm}\right) + 0.0681\right]$$
$$(8.13)$$

Eq. 8.13 is used to get the energy absorbed by CO_2 in the atmosphere. A spreadsheet can be developed to calculate the average earth temperature as a function of the CO_2 in the atmosphere. **Table 8.2** is the first few lines of the spreadsheet.

TABLE 8.2 Extrapolating Keeling Curve to find average earth temperature.

Year	Cppm	Moles/m² (4-11)	Absorption Fraction (4-9)	Absorbed by CO_2 (4-12)	Other GHG (Wt/m²)	Total Absorbed (Wt/m²)	Emitted (Wt/m²)	Temperature (°K)	Temperature Difference (°K)
2000	368	128.4	0.5091	48.55	102.45	151	390	287.985	0
2001	370.2	129.2	0.5097	48.60	102.45	151.05	390.05	287.995	0.010
2002	372.4	130.0	0.5102	48.66	102.45	151.11	390.11	288.006	0.020
2003	374.6	130.7	0.5113	48.71	102.45	151.17	390.17	288.016	0.031
2004	376.8	131.5	0.5113	48.77	102.45	151.23	390.23	288.027	0.042
2005	379	132.3	0.5118	48.83	102.45	151.28	390.28	288.038	0.052
2006	381.2	133.0	0.5123	48.89	102.45	151.34	390.34	288.048	0.063
2007	383.4	133.8	0.5128	48.94	102.45	151.40	390.40	288.059	0.073

© SAE International

FIGURE 8.4   Global temperature increase.

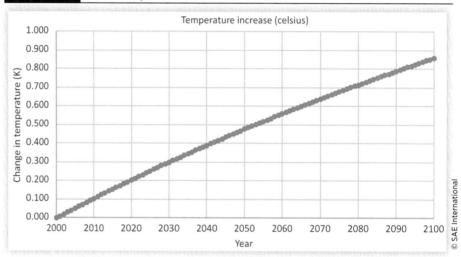

© SAE International

A plot of the results is shown in **Figure 8.4**. This model shows a temperature increase of 0.86°K over 100 years or about 1.5°F. The model is an oversimplification of what actually happens, because it assumes the earth is of constant temperature and that the radiation it receives and emits is constant over the surface of the earth. It is easy to find faults with the model. But this model includes the main effects of carbon dioxide absorbing infrared radiation and how it increases the temperature of the surface of the earth. Climate scientists have developed much more sophisticated models, but the results are similar to the results of this model.

References

1. https://en.wikipedia.org/wiki/Solar_irradiance
2. https://www.ess.uci.edu/~yu/class/ess200a/lecture.2.global.pdf
3. Stull, Wyatt and Plass, *Applied Optics* 3(2):250, 1964.
4. https://en.wikipedia.org/wiki/Keeling_Curve

Fundamentals of Batteries

9.1 Introduction

Before reading through this unit I think it is helpful to watch the NOVA movie "Search for the Super Battery". It is a little outdated, but there is good information in the movie about batteries. As we move toward having electric and hybrid electric cars, the battery system is going to be very important.

The ideal battery would be able to store and provide energy at any rate. That is, we would be able to charge and discharge the battery as slowly or as quickly as we want. The battery would be 100% efficient when charging and discharging. None of the electric energy would be converted to heat. The battery would be small and light weight.

Real batteries are nothing like the ideal battery, as described earlier. We must be careful how fast we charge and discharge batteries to prevent them from damaging. It takes hours to fully charge a battery, whereas, it takes only a few minutes to fill a car with gas. People who own electric cars do not want to wait hours for the batteries to charge.

Real batteries are heavy and take up a considerable amount of space in a vehicle. Batteries have good efficiency, but some of the energy is always converted to heat when charging and discharging a battery. A good battery system is about 80% efficient in typical use, meaning that about 80% of the electric energy used to charge the battery system will be recovered when the battery system is discharged. The efficiency is rate sensitive. Charging and/or discharging the battery system quickly forces it to operate at a lower efficiency.

Batteries can be classified as two types: primary batteries and secondary batteries. Primary batteries are the small batteries that are used around the house. They are discharged once and then thrown away. Primary batteries are not useful for electric vehicles and will not be discussed. Secondary batteries are rechargeable and are the type of batteries that are useful for electric vehicles. There are many types of secondary batteries, lithium-ion, lead-acid, nickel-cadmium, and nickel-metal-hydride are most common. There are other secondary battery types that have specialized uses.

Common battery terminology is as follows:

- Energy Density: The amount of energy that can be stored in a battery divided by its mass (W-h/kg).

- Amp-Hour Capacity: The amount of charge the battery will hold in amp-h (amp-h = 3600 C).

- Watt-Hour Capacity: The amount of energy the battery will hold in W-h. The watt-hour capacity of a battery depends on the rate the battery is discharged. Discharging the battery at a high rate will cause the energy to be drawn out at a lower voltage, which reduces the amount of energy that can be drawn out of the battery.

- C Rating: This is the charge or discharge rate for the battery that was used to measure the amp-hour and watt-hour capacity. A rate of C/3 means the battery was charged or discharged completely in 3 h. C/8 means it was charged or discharged in 8 h, etc.

Batteries are very heavy and take up a lot of space for the amount of energy they store. **Figure 9.1** shows the energy density of several energy sources.

FIGURE 9.1 Energy density of common energy sources.

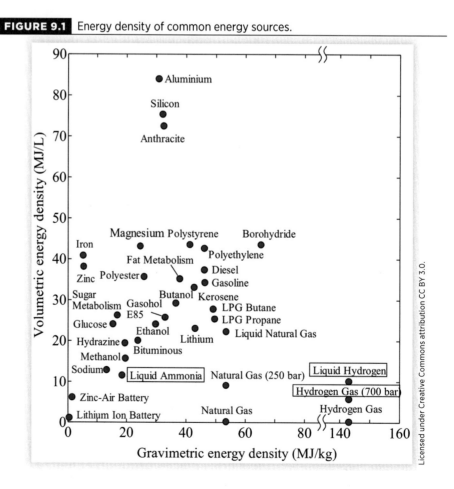

The x-axis in Figure 9.4 is the energy density associated with mass (MJ/kg). The y-axis is the energy density associated with volume (MJ/L). Ideally an energy source would provide a lot of energy for the mass and the amount of volume it occupies. The perfect energy source would be in the top right-hand corner of the chart. No such perfect energy source exists. Hydrogen gas has the highest energy content for its mass, but it takes up a lot of volume on the vehicle. Burning aluminum metal yields the highest amount of energy for the volume it occupies. The metals aluminum, silicon, iron, zinc, and magnesium are not practical fuels for most applications. The energy to make these elements as a fuel would exceed the energy recovered when they are burned.

The lithium-ion battery is the highest energy density battery that we currently have for practical use. The zinc-air battery has a higher energy density than lithium-ion but is impractical except for a few specialty applications. Notice the energy density of lithium-ion batteries is extremely low compared to any of the fossil fuels we commonly use. Lead-acid batteries are even lower than lithium-ion both in energy per kilogram and energy per liter. Batteries will be much heavier and occupy much more space than any of the other fuel sources we use to power vehicles. This is one of the fundamental problems of using batteries to power vehicles.

9.2 **Battery Efficiency**

There are two types of battery efficiency, the charge efficiency and the energy efficiency. The charge efficiency is the amp-h charge recovered from the battery when it is discharged divided by the amp-h required to charge the battery.

$$\text{Charge Efficiency} = \frac{\text{Amp-h recovered discharging battery}}{\text{Amp-h required to charge battery}} \qquad (9.1)$$

Charge efficiency for the batteries used in electric vehicles is usually very high, almost always above 90% and often very near 100%. Batteries that have a liquid electrolyte, like lead-acid batteries, will use a small amount of charge in the electrolysis of water, especially when charging the battery. Charge is a fundamental quantity and must be conserved; it is a basic law of physics. But, if there is electrolysis of water, the test results will indicate that more amp-h was put into charging the battery than was released when discharging and the charge efficiency will be less than 100%. There are other secondary effects that can cause the charge efficiency to be less than 100%, and batteries will have a charge efficiency less than 100%, though it will be near 100% for batteries we are interested in.

Energy efficiency is the watt-h energy we get discharging a battery divided by the watt-h energy required to charge the battery.

$$\text{Energy Efficiency} = \frac{\text{Watt-h recovered discharging battery}}{\text{Watt-h required to charge battery}} \qquad (9.2)$$

When charging and discharging a battery, some of the energy is converted to heat. This manifests itself as a voltage difference. We charge the battery at a higher voltage and discharge the battery at a lower voltage. This is illustrated in **Figure 9.2:**

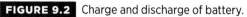

FIGURE 9.2   Charge and discharge of battery.

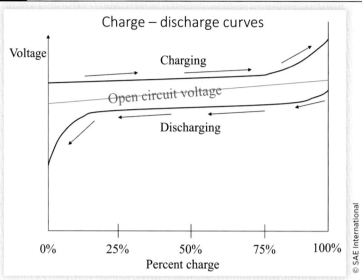

Plotted on the *x*-axis in Figure 9.2 is the percent charge in the battery such that at 0% the battery is completely discharged and at 100% the battery is completely charged. The blue line on the chart represents the open circuit voltage. If we use a high-quality voltmeter and measure the voltage across the terminals at different states of charge we will develop the open circuit voltage curve. The open circuit voltage is drawn as a line in Figure 9.2, though it will not be exactly linear, and the exact curve depends on the type of battery. Open circuit voltage is the equilibrium voltage for the battery. Charge will not flow in or out of the battery if the voltage across the terminals is equal to the open circuit voltage. The open circuit voltage is always lower at low states of charge and higher at high states of charge. Figure 9.2 shows the open circuit voltage to vary linearly with charge, but it will not be exactly linear, though it is approximately linear except for the portions near 0% and 100% charge.

To put charge into the battery we must increase the voltage across the terminal above the open circuit voltage. A charging path is illustrated in Figure 9.2 above the open circuit voltage curve. The voltage across the terminals is held above the open circuit voltage and charge flows into the battery until it is charged. Most batteries will see the voltage curve rise significantly as the battery approaches full charge as illustrated in Figure 9.2. We will need to reduce the charging rate as the battery approaches full charge to avoid over-voltage and damaging the battery.

In order to get charge to flow out of the battery we must decrease the voltage across the terminals to a value below the open circuit voltage. A discharge path is illustrated in Figure 9.2 below the open circuit voltage curve. When discharging a battery, the voltage will drop significantly as we approach 0% charge in the battery, as illustrated in Figure 9.2. We will need to reduce the rate of discharge as we approach 0% charge to prevent under-voltage and damage to the battery.

Energy is voltage multiplied by charge, i.e., Joule = Volt × Coulomb. The area under the charging curve in the figure represents the energy required to charge the battery and the area under the discharge curve represents the energy we get back from the battery when discharging. It always takes more energy to charge the battery than what we get back when discharging. The battery is always less than 100% efficient in storing and providing energy.

Let V_{Charge} be the average voltage used in charging the battery and $V_{Discharge}$ be the average voltage when discharging the battery. If the battery had 100% charge efficiency, then the energy efficiency would be the ratio of the two voltages. As discussed earlier, because of electrolysis of water and other effects, testing of the battery will indicate a charge loss. That is, it will take more amp-h to charge the battery than what we get back in discharging the battery. The batteries we are interested in will have a high charge efficiency, but it will be less than 100%. The energy efficiency of the battery is expressed as follows:

$$\text{Energy Efficiency} = \left(\frac{V_{Discharge}}{V_{Charge}}\right)(\text{Charge Efficiency}) \qquad (9.3)$$

The battery efficiency is low at low states of charge because the discharge voltage is low. The battery efficiency is low at high states of charge because the charging voltage is high. Generally, the battery is most efficient when operated between 25% and 75% state of charge. To get the highest energy efficiency from the battery we should keep it operating in the middle range as far as state of charge.

9.2.1 Charging the Battery

To drive current into the battery the voltage across the terminals must be higher than the open circuit voltage. Increasing the rate of charge flowing into the batteries requires higher voltage, but there is a limit to how high we can push the voltage without damaging the battery. Charging and discharging a battery causes a chemical reaction inside the battery involving the positive and negative plates and the electrolyte. The chemical reaction is reversible, going in one direction for charging and in the other direction for discharging. Applying too high of a voltage across the battery terminal will cause undesirable chemical reactions on the plates, which damages the battery. Macroscopically we specify a maximum voltage that can be applied to the battery terminals to avoid damaging the batteries.

As the battery approaches full charge most of the area on the positive and negative plates has gone through the chemical reaction, and there is a limited amount of active plate area left. The charging rate will need to be reduced as the batteries approach full charge to avoid over-voltage on the batteries. All types of batteries require that the charging rate be reduced as they approach full charge.

9.2.2 Discharging the Battery

To discharge the battery, we must reduce the voltage across the terminals below the open circuit voltage. Increasing the rate of discharge requires that the charge be drawn out at lower voltage, but there is a limit to how fast we can discharge the battery. If we discharge too fast we will cause undesirable chemical reactions on the plates that damage the battery. Since energy is voltage multiplied by charge, it is also true that discharging at

FIGURE 9.3 Effect of charging rate on battery efficiency.

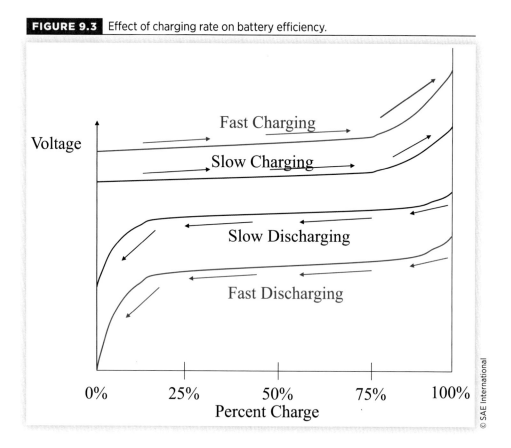

© SAE International

low voltages means we are not getting much energy from the battery. It is undesirable to discharge the battery at a low voltage, and it can damage the battery.

As the battery approaches complete discharge there is a limited amount of area on the plates that can continue the chemical reaction. The discharge rate needs to reduce as we approach complete discharge to avoid damaging the battery and to get a significant amount of energy from the battery.

Figure 9.3 illustrates how charging and discharging a battery quickly forces the battery to operate at a lower efficiency. For electric vehicles we want to be able to charge the batteries quickly so we can get back on the road. But this requires that we use a high charging voltage, which causes the batteries to operate at a lower efficiency. It will take more energy to charge the batteries quickly than to charge them slowly.

Figure 9.3 is a schematic that illustrates how charging rate affects the efficiency of a battery. Charging and discharging at a faster rate requires a higher charge voltage and a lower discharge voltage, which reduces the efficiency of the battery. When fast charging and discharging, a higher percentage of the energy going in and out of the battery will be converted to heat, and the battery will heat up. Too much heat will damage the battery. Charging at high voltage will damage the battery and discharging at low voltage will damage the battery. Charge and discharge rates are limited to protect the battery. Trying to charge or discharge the battery too quickly will damage the battery.

Figure 9.4 is a typical charge and discharge curve for a lithium-ion battery used in a solar car project at Missouri S&T. Notice how the charging rate was reduced as the battery approached full charge. The maximum charge voltage used on this cell was 4.2 V.

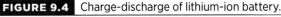

FIGURE 9.4   Charge-discharge of lithium-ion battery.

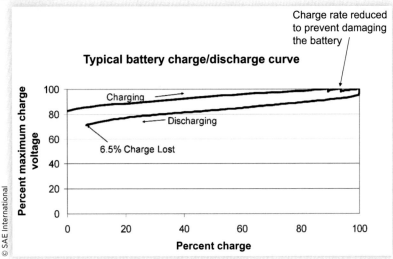

The tester was programmed to cut the charge rate current in half when the voltage approached 4.2 V. This battery also shows a 6.5% charge loss on discharge, which is what the test would show on the first cycle. When running charge-discharge cycles on a battery the discharge voltage is limited to a value that will not damage the battery. In the first cycle it is likely that the cutoff voltage on discharge caused the battery to stop before all of the charge was drawn out of the battery. The charge efficiency will improve with the second cycle. Figure 9.4 is typical for the first cycle of the test.

9.2.3 Measuring State of Charge

The open circuit voltage of a battery varies with state of charge, so measuring open circuit voltage would be a good way to measure the state of charge of a battery. The problem with this method is that to get a true open circuit voltage measurement the battery must stabilize at room temperature. If the battery is being charged or discharged it takes a while for the battery chemistry to stabilize enough to take an accurate open circuit voltage reading. The battery needs to sit for at least an hour to stabilize, and we often do not want to wait an hour to know the state of charge of the battery. There are amp-h meters that keep track of the charge going into and out of the battery. The amp-h meters are a much more accurate and reliable way to measure the amount of charge in the battery.

9.3 Battery Capacity and Efficiency

Gibbs free energy: The Gibbs free energy is based on the chemical potential of the battery chemistry and is the maximum amount of energy that can be drawn out of the battery assuming 100% efficiency. The Gibbs free energy formula is:

$$\Delta G = -n\,F\,V \tag{9.4}$$

In the formula n is the number of electrons involved in the chemical reaction. F is the Faraday constant, which is 96,485 coulombs/mole, and V is the reversible (open circuit) voltage. ΔG is the amount of energy that can be drawn out of the battery, known as the Gibbs free energy.

Example 9.1: The best way to illustrate how the Gibbs free energy formula can be used to estimate battery capacity is to work through an example. The copper-zinc battery was one of the first batteries developed. There are no practical applications for the battery today, but the chemistry is simple enough that it is a good place to start talking about the specifics of battery chemistry. This example calculates the Gibbs free energy for the copper-zinc battery.

Solution: When measuring chemical potential, we can only measure the difference in chemical potential between two materials. As a reference we define the chemical potential of hydrogen as zero. For hydrogen, the chemical potential $V^0 = 0.0$ V by definition. Based on that definition the chemical potential of copper is $V^0 = 0.337$ V and that of zinc is $V^0 = -0.763$ V. There are readily available tables of chemical potential values for many different materials.

The reversible (open circuit) voltage for a copper-zinc battery is the difference of the chemical potentials:

$$V = (0.337\,\text{V}) - (-0.763\,\text{V}) = 1.1\,\text{V} \qquad (9.5)$$

The chemical reaction is illustrated in **Figure 9.5**:

FIGURE 9.5 Copper-zinc battery.

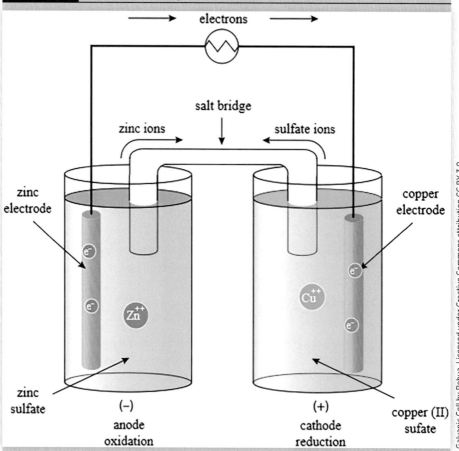

The zinc and copper cylinders in Figure 9.5 represent the zinc and copper plates in the battery. The zinc plates are in an electrolyte solution of $ZnSO_4$ and the copper plates are in an electrolyte solution of $CuSO_4$, as illustrated in the figure. There is a porous separator between the electrolytes that allows SO^{-2} ions to pass through but blocks the copper and zinc ions.

When discharging, the zinc metal goes into solution as Zn^{+2} ions, leaving 2 electrons in the zinc plate. To balance the charge, two electrons go up the wire, through the light bulb, and across toward to the copper plate. The Cu^{+2} ion comes out of solution and is deposited on the copper plate. To balance the charge the copper ion needs the two electrons that are coming across the wire from the zinc plate. During the reaction the zinc plate gradually dissolves into solution and the copper plate grows as copper is added to the plate.

Charging the battery causes the chemical reaction to go the other direction with zinc being deposited on the zinc plate and copper going into solution. There are two electrons involved in the battery chemistry, so n = 2 in the Gibbs free energy equation. The Gibbs free energy for this battery can be calculated from Eq. 9.4 as:

$$\Delta G = -(2)(96,485)(1.1) = 212,267 \, J / mole \tag{9.6}$$

The result is that the battery stores 212,267 J of energy for each mole of material involved in the chemical reaction. One mole of zinc has a mass of 59 g and one mole of copper has a mass of 65 g. A copper-zinc battery must have a mass of 59 + 65 = 124 g of active copper and zinc to provide 212,267 J of energy. As an upper limit, the energy density of a copper-zinc battery is:

$$\text{Upper Limit Energy Density} = \frac{212,267 \, J}{124 \, g} = 1.7 \, MJ / kg \tag{9.7}$$

Achieving this energy density would require that the electrolyte and other parts of the battery have a zero mass, which is not possible. The energy density of a copper-zinc battery will always be less than 1.7 MJ/kg. If a battery could be developed so that most of the mass was in the active copper and zinc materials, then the energy density of the battery could approach 1.7 MJ/kg. All real copper-zinc batteries have an energy density less than 1 MJ/kg. For comparison, diesel fuel and jet fuel have an energy density of about 47 MJ/kg. So even under the most ideal conditions a petroleum fuel has 25+ times more energy per kg than a copper-zinc battery.

This type of analysis can be done for any battery chemistry. We examine the mass of the active materials and can compute an upper limit for the energy density. Having active materials with a low atomic mass will allow us to make batteries that have a higher energy density. Hydrogen is element number 1 and has the lowest atomic mass. A fuel cell is a lot like a battery that uses hydrogen as one of the active materials. Fuel cells have a higher energy density than batteries, but the energy density of a fuel cell is much lower than that of petroleum fuel. Helium is element number 2, but helium is chemically inert, so it is impossible to make a helium battery. Lithium is element number 3 and offers the best potential for making a lightweight battery. The lithium-ion battery has a high energy density because it uses lithium as the active material.

Battery Voltage: The open circuit voltage of a battery is obtained by looking up the chemical potential V^0 of the products. For the copper-zinc battery the copper has a chemical potential of 0.337 V and the zinc's is -0.763 V. The open circuit voltage was calculated as 1.1 V. When we begin charging or discharging the battery the voltage across the terminals will change because of the kinetics of the chemical reaction. The change in voltage depends on temperature and the activities of the products and reactants in the chemical reaction. The equation is normally written as:

$$V_r = V_0 - \left(\frac{R\,T}{n\,F}\right)\ln\left(\frac{\text{activities of products}}{\text{activities of reactants}}\right) \tag{9.8}$$

In Eq. 9.8, V_0 is the open circuit voltage, R is the constant for the ideal gas law, T is absolute temperature, n is the number of electrons in the chemical reaction and F is the Faraday constant. It is almost impossible to get accurate values for the activities of the products and reactants, and the values vary with state of charge, so Eq. 9.8 has limited practical value. The change in voltage that happens as we charge or discharge the battery is described chemically as activation or charge transfer over-potential and concentration or mass transfer over-potential, which are described later.

Activation or Charge Transfer Over-potential: As the chemical reaction proceeds to fully charged or fully discharged the active plate area decreases. The reduction in active area on the plates will cause a cell over-potential because it becomes more difficult to find areas on the plates where the chemical reaction can proceed. For example, if we are charging a battery from a low state of charge, the chemical reaction can proceed almost anywhere on the plates. As the battery approaches full charge most of the plates have gone through the chemical reaction and there is a limited number of places on the plates for the chemical reaction to proceed. The rate of the chemical reaction will slow down, and we will measure a reduced current going into the battery. Increasing the charging voltage will cause the chemical reaction to proceed faster. If the voltage is raised too high it will cause undesirable chemical reactions to be deposited on the plates, effectively reducing the amount of active material in the battery. The undesirable chemical reactions manifest themselves as lower battery capacity, and, in most cases, it is impossible to recover the capacity that is lost. This is why it is important to not over-voltage the battery.

Concentration or Mass Transfer Over-potential: The chemical reaction uses up ions in the electrolyte. As the chemical reaction proceeds, the ion concentration near the plates will decline as ions near the plates are absorbed and used up. This causes the ion concentration near the plates to be lower than the average ion concentration in the electrolyte. The ions must physically move through the electrolyte to participate in the chemical reaction at the plates, and this does not happen instantaneously. Macroscopically this will manifest itself as a higher charging voltage or lower discharge voltage.

Internal Resistance: The cell over-potential is approximately proportional to the amount of current charging or discharging from the battery. Macroscopically, it appears that the battery has an internal resistance, though we must be careful when we model the internal resistance of the battery. Please remember that it is not physically an electrical resistance, but a way to model the chemical reaction of the battery. The internal resistance will decrease the discharge voltage and increase the charge voltage. If you simply put a resistor in series with the battery it gives the wrong answer. The resistor needs to be inside the battery between the two terminals for charging the battery and outside the terminals when discharging the battery. If the internal resistance of the battery is R and the current is I:

$$V_{discharge} = V_0 - IR \qquad (9.9)$$

$$V_{charge} = V_0 + IR \qquad (9.10)$$

The energy efficiency of the battery can be expressed as:

$$\text{Energy Efficiency} = \left(\text{Charge Efficiency}\right)\left(\frac{V_0 - IR}{V_0 + IR}\right) \qquad (9.11)$$

Example 9.2: Assume a 60 amp-hr deep cycle lead-acid battery with an internal resistance of 0.07 Ω, and open circuit voltage of 12.6 V and a charge efficiency of 97%. Calculate the energy efficiency for charging the battery at 10 amps and discharging the battery at 10 amps.

Solution: Using Eq. 9.11:

$$\text{Energy Efficiency} = \left(0.97\right)\left(\frac{12.6 - \left(10\right)\left(0.07\right)}{12.6 + \left(10\right)\left(0.07\right)}\right) = 86.8\% \qquad (9.12)$$

If we charge the battery at 10 amps and discharge at 10 amps we will recover 86.8% of the energy. That is, if we put 100 W-h of energy into charging the battery, we will recover 86.8 W-h when discharging. The remaining 13.2 W-h of energy will be converted to heat during the charge/discharge cycle.

Example 9.3: For the 60 amp-h battery above it would take about 6 h to charge at 10 amps. Suppose we want to cut the charging time two hours by charging at 30 amps rather than 10 amps. We will still discharge the battery at 10 amps. How does this affect the energy efficiency of the battery?

Solution: Use Eq. 9.11:

$$\text{Energy Efficiency} = \left(0.97\right)\left(\frac{12.6 - \left(10\right)\left(0.07\right)}{12.6 + \left(30\right)\left(0.07\right)}\right) = 78.5\% \qquad (9.13)$$

This example illustrates how charging the battery faster forces it to operate at a lower efficiency. Charging faster causes a higher portion of the energy to be converted to heat, and the battery will heat up. Too much heat can damage the battery. For electric vehicles it is desirable to charge the batteries quickly but charging quickly will waste more of the electric energy used to charge the battery and it will be more expensive to charge the battery. Wasting energy to charge the batteries takes away some of the benefits of driving electric cars.

Figure 9.6 was developed for the 60 amp-h battery in Examples 9.2 and 9.3, assuming that the charging and discharging amperage is the same. The chart shows how energy efficiency of the battery varies with the amperage used in charging and discharging the battery.

FIGURE 9.6 Battery efficiency vs amperage.

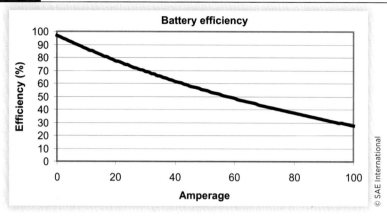

9.4 **Battery Chemistry**

9.4.1 **Lead-acid Batteries**

Lead-acid batteries are established technology. They are low cost, reliable, and durable. There are many practical applications for lead-acid batteries. The starter batteries in cars and trucks are almost always lead-acid batteries. Lead acid batteries are used in wheelchairs, trolling motors, forklifts, golf carts and other small electric vehicles. It is the most common rechargeable battery used. Lead-acid batteries are too heavy to be used to power electric cars and trucks, but they could be used for grid storage. **Figure 9.7** shows a battery room in a power plant with racks of lead-acid batteries.

FIGURE 9.7 Battery room in power plant.

The active materials in a lead-acid battery are lead on the negative plates, lead oxide on the positive plates and sulfate ions in the electrolyte. When discharging the battery, the following chemical reaction happens on the negative plates.

$$\text{Negative Plates}: Pb + SO_4^{-2} \rightarrow PbSO_4 + 2e^- \qquad (9.14)$$

Discharging the battery causes the negative plates to become coated with lead sulfate. The electrons on the right side of Eq. 9.14 become the current the battery produces. Charging the battery causes the chemical reaction to go in the other direction. The sulfate ions go back into solution and the negative plates go back to being coated with lead.

The positive plates start out being coated with lead oxide. Discharging the battery causes the following chemical reaction on the positive plates.

$$\text{Positive Plates}: PbO_2 + 4H^+ + SO_4^{-2} + 2e^- \rightarrow PbSO_4 + 2H_2O \qquad (9.15)$$

The chemical reaction on the positive plates is more complex than on the negative plates. Four hydrogen ions and one sulfate ion must come in close proximity to the lead oxide molecule for the chemical reaction to take place. The oxygen atoms in the lead oxide must be stripped off to combine with the hydrogen ions and make water. The sulfate ion combines with the lead to make lead sulfate. As the reaction proceeds, the positive plates are coated with lead sulfate. There is an excess charge of positive 2 on the left side that is balanced by the 2 electrons coming from the negative plates. The total chemical reaction can be written as:

$$\text{Total Reaction}: Pb + PbO_2 + 2H_2SO_4 \rightarrow 2PbSO_4 + 2H_2O \qquad (9.16)$$

As the battery is discharged the positive and negative plates are coated with lead sulfate. The sulfate ion concentration in the electrolyte decreases as the lead sulfate is deposited on the plates. The density of the electrolyte will decrease as the battery is discharged because of the loss of sulfate ions. It is possible to measure state of charge in a lead-acid battery by measuring the density of the electrolyte. A schematic of the lead-acid battery cell is illustrated in **Figure 9.8.**

FIGURE 9.8 Lead-acid battery schematic.

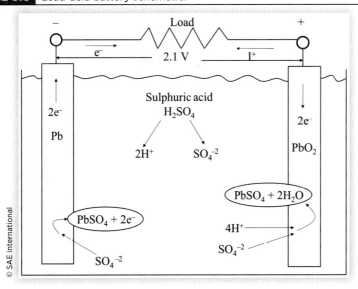

© SAE International

As the battery nears complete discharge, the plates are nearly completely coated with lead sulfate ($PbSO_4$). The reaction can proceed only when four free hydrogen ions and one free sulfate ion come together at an uncovered spot on the PbO_2 plate. From a probability standpoint, this becomes less likely to happen, and the chemical reaction must slow down. There is still a significant amount of energy that can be drawn out of the battery, but it must be drawn out slowly. Trying to drive the reaction too fast leads to $PbO \cdot PbSO_4$ and other undesirable chemical compounds on the positive plate. Some of the chemical reactions are almost irreversible, and permanently degrade the energy storage capacity of the battery by partially coating the plates.

As the battery nears complete charge there is very little lead sulfate ($PbSO_4$) left on the surface. Driving the chemical reaction too fast will cause adverse chemical compounds to form on the plate surface, which reduces the battery capacity. Charging and discharging rates must be reduced as the battery approaches full charge or full discharge. This is true for all batteries.

- When charging or discharging at a fast rate, ions must move through the electrolyte at a high rate.
- The electrolyte near the plates becomes saturated with ions moving out of the plates and has low concentration of ions moving into the plates.
- The saturation and low concentrations tend to drive the chemical reaction in the opposite direction.
- This gives an apparent increase in electrical resistance.

Each cell in a lead acid battery has a voltage of nominally 2.1 V when fully charged. Most lead-acid batteries have six cells wired in series. They have a nominal voltage of 12.6 V when fully charged and are referred to as 12 V batteries. The automotive industry settled on a 12 V system, and because of that, there are many lights, motors, and other electrical devices designed to operate at 12 V. The abundance of electrical devices available makes a 12 V system popular for low power applications. Lead-acid batteries can be produced at any voltage that is a multiple of the 2.1 V per cell, and 6 V batteries are common. For higher operating voltages it is common to put 12 V batteries in series. Forklifts are commonly powered by 24 V, 36 V, and 48 V lead-acid batteries.

Weakest Cell: Battery cells are put in series to generate a working voltage. There is always a weakest cell, which is the cell that has the lowest energy storage capacity. The weakest cell will always reach complete charge and discharge first when charging and discharging the battery. The weakest cell limits the capacity of the entire battery. Overcharging or discharging of the weakest cell will further reduce its capacity, which further degrades the battery. When a battery fails because its capacity is too low to be useful, it is often really just one cell that has failed, and the rest of the cells are good.

When charging a lead-acid battery some of the electrons going into the negative plates will combine with hydrogen ions in the electrolyte to form hydrogen gas. Some of the electrons being released into the negative plates will come from oxygen ions in solution which will cause the release of oxygen gas near the positive plates. **Figure 9.9** illustrates the electrolysis of water in a lead-acid battery.

Electrolysis of Water: For batteries that have a water-based electrolyte like sulfuric acid or potassium hydroxide there is always some electrolysis of water when charging

FIGURE 9.9 Electrolysis of water in lead-acid battery.

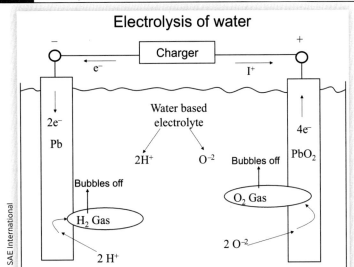

and discharging the battery. This electrolysis of water gives an apparent charge loss when charging and discharging the battery. It will take more amp-h to charge the battery than will be recovered when discharging the battery. There is always some charge leakage in batteries, but most of the charge loss goes to electrolysis of water. An excessive amount of electrolysis will deplete the electrolyte. Older lead-acid batteries required that water be added to the electrolyte occasionally. The newer lead-acid batteries have been designed so that it is not normally necessary to add water.

9.4.2 Silver-Zinc Batteries

Silver-zinc batteries were one of the first batteries developed. They are still being used for military applications because the electrolyte can be stored separately and then added to the battery when the battery is needed. This gives the silver-zinc battery a long shelf life which makes it suitable for certain military applications. These batteries are expensive because of the silver on the positive plates.

Once the electrolyte is added the batteries have a short life compared to lead-acid or lithium-ion batteries. Some silver-zinc batteries can only be charged and discharged 10 or 20 times before they are worn out.

The negative plates in a silver-zinc battery are coated with zinc and the positive plates are coated with silver oxide. The electrolyte is a potassium hydroxide solution. **Figure 9.10** illustrates the chemistry of the silver-zinc battery.

The chemical reaction on the positive plate is complex. It can be written in three steps as:

$$2K^+ + AgO + 2e^- \rightarrow K_2O + Ag \qquad (9.17)$$

$$K_2O + Ag \rightarrow 2K^+ + O^{-2} + Ag \qquad (9.18)$$

FIGURE 9.10 Silver-zinc battery schematic.

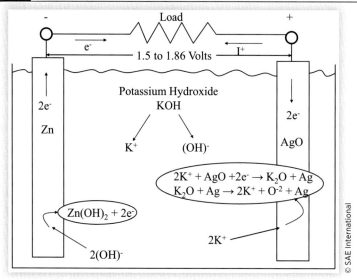

$$O^{-2} + 2H^+ \rightarrow H_2O \tag{9.19}$$

The chemical reaction on the negative plate is:

$$Zn + 2\left(OH\right)^- \rightarrow Zn\left(OH\right)_2 + 2e^- \tag{9.20}$$

The net chemical reaction is:

$$Zn + AgO + H_2O \rightarrow Zn\left(OH\right)_2 + Ag \tag{9.21}$$

9.4.3 Nickel-Cadmium Batteries

Nickel-cadmium batteries have been used in many applications and there are many sizes of batteries. They have an energy density that is higher than lead-acid, but lower than lithium-ion. They are reliable and can be cycled many times. The nickel-cadmium battery uses cadmium as the negative plate and the positive plates are coated with nickel oxyhydroxide. The electrolyte is potassium hydroxide. A schematic illustrating the battery chemistry is shown in **Figure 9.11**. The chemistry is similar to the silver-zinc battery.

The chemical reaction on the positive plate is complex. It can be written in three steps as:

$$2K^+ + OH^- + NiO\left(OH\right) + e^- \rightarrow K_2O + Ni\left(OH\right)_2 \tag{9.22}$$

$$K_2O + Ni\left(OH\right)_2 \rightarrow 2K^+ + O^{-2} + Ni\left(OH\right)_2 \tag{9.23}$$

$$O^{-2} + 2H^+ \rightarrow H_2O \tag{9.24}$$

FIGURE 9.11 Nickel-cadmium battery schematic.

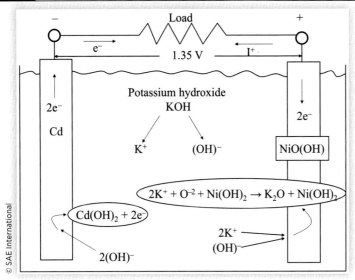

The chemical reaction on the negative plate is:

$$Cd + 2(OH)^- \rightarrow Cd(OH)_2 + 2e^- \tag{9.25}$$

The net chemical reaction is:

$$Cd + 2NiO(OH) + 2H_2O \rightarrow Cd(OH)_2 + 2Ni(OH)_2 \tag{9.26}$$

9.4.4 Nickel-Hydrogen Batteries

Nickel-hydrogen batteries have been used in satellites. The energy density is significantly higher than for lead-acid batteries, but lower than for lithium-ion batteries. The batteries are capable of many charge/discharge cycles, which is important for space applications. A pressure vessel is required to store the hydrogen gas.

The hydrogen gas is the negative plate in the battery, so a porous electrode must be developed to allow the hydrogen gas to flow in and out of the negative plate. The hydrogen gas is used up when the battery is discharged and replenished when the battery is charged. The pressure inside the battery can be used to indicate the state of charge. The positive plates in the battery are nickel oxy-hydroxide and the electrolyte is potassium hydroxide. The chemistry is similar to the nickel-cadmium battery. **Figure 9.12** is a schematic of the nickel-hydrogen battery.

The chemical reaction on the positive plate is complex. It can be written in three steps as:

$$2K^+ + OH^- + NiO(OH) + e^- \rightarrow K_2O + Ni(OH)_2 \tag{9.27}$$

$$K_2O + Ni(OH)_2 \rightarrow 2K^+ + O^{-2} + Ni(OH)_2 \tag{9.28}$$

$$O^{-2} + 2H^+ \rightarrow H_2O \tag{9.29}$$

FIGURE 9.12 Nickel-hydrogen battery schematic.

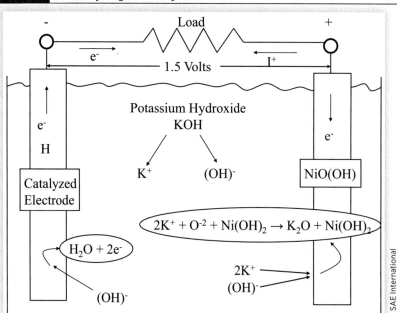

The chemical reaction on the negative plate is:

$$H_2 + 2(OH)^- \rightarrow 2H_2O + 2e^- \tag{9.30}$$

The net chemical reaction is:

$$H_2 + 2NiO(OH) \rightarrow 2Ni(OH)_2 \tag{9.31}$$

9.4.5 Nickel Metal Hydride Batteries

Nickel metal hydride batteries are used in many applications. They are reliable and durable and have a higher energy density than lead-acid batteries. The chemistry is the same as the nickel-hydrogen batteries, but the hydrogen is stored in a metal hydride rather than in a pressure vessel. The catalyzed electrode in the previous figure is replaced with a metal hydride electrode.

9.4.6 Lithium Batteries

Lithium is a lightweight element and people have tried for many years to make lithium batteries because of the potential for lightweight batteries. Lithium is very reactive and explosive, and batteries made using lithium metal have always been regarded as too dangerous to be used in practice. A lot of interesting work has been done, but no practical batteries have been developed.

9.4.7 Lithium-Ion Batteries

Lithium-ion batteries are very different from other battery chemistries. There is no metallic lithium in the batteries. The charge moves in the batteries by transporting

FIGURE 9.13   Graphite structure.

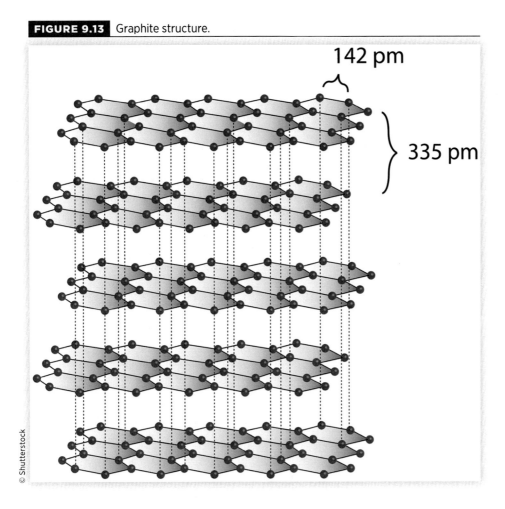

142 pm

335 pm

© Shutterstock

lithium-ions across an organic or polymer membrane electrolyte. Lithium is a small enough atom to fit into the crystal lattice of larger atoms. The negative electrode is made of carbon graphite, which is a layered material and the lithium atoms fit between the layers in the graphite. The structure of graphite is illustrated in the **Figure 9.13**:

Because of the way the lithium atoms fit into the graphite structure there need to be 6 carbon atoms for each lithium atom. That is the maximum lithium that can be put into the graphite negative terminal of the battery.

The positive terminal is made from a metal oxide. The earliest lithium-ion batteries used cobalt oxide for the positive. Nickel oxide and manganese oxide have been used, and a blend of metal oxides has been used. The chemical reaction for lithium-ion batteries is shown assuming cobalt oxide is used for the positive terminal. The positive plates follow the chemical reaction:

$$CoO_2 + Li^+ + e^- \rightarrow LiCoO_2 \tag{9.32}$$

The negative plate reaction is:

$$LiC_6 \rightarrow C_6 + Li^+ + e^- \tag{9.33}$$

FIGURE 9.14 Schematic of lithium-ion battery.

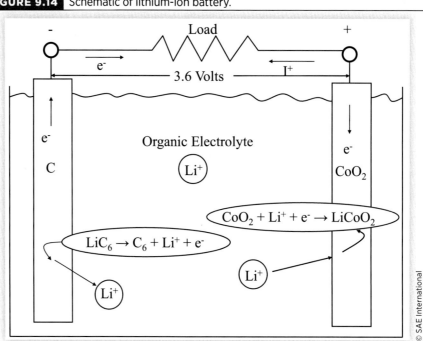

FIGURE 9.14 Schematic of lithium-ion battery.

The overall reaction is:

$$LiC_6 + CoO_2 \rightarrow C_6 + LiCoO_2 \tag{9.34}$$

A schematic of the lithium-ion battery is illustrated in the figure below (**Figure 9.14**):

The lithium-ions move across the organic or polymer electrolyte. The electrolyte should be very thin to reduce internal resistance in the battery. The early lithium batteries had problems with dendritic growth through the electrolyte. As the lithium-ions moved back and forth between the plates during the charge/discharge cycles an occasional bridge of lithium metal would connect the positive and negative plates shorting the battery out. The result was a catastrophic failure and the battery would sometime ignite. Kyocera had a major recall of products in the early 2000s because of this type of battery failure. The more recent lithium batteries have been designed in such a way as to prevent the bridging and shorting out and are much safer.

9.5 Homework

1. A 5-kW-h, lithium-ion battery system has a weight of 80 lb. A 5-kW-h, lead-acid battery system has a weight of 300lb.

    a. Find the energy density of the two battery systems in kilowatt-hours per kilogram (KWH/kg) and in mega-joules per kilogram (MJ/kg).

    b. Use the Figure 9.1 to compare the energy density of gasoline to the energy density of the two battery systems. (1) How many kg of lithium-ion batteries would it take to provide the same amount of energy as one kg of gasoline? (2)

How many kg of lead-acid batteries would it take to provide the same amount of energy as one kg of gasoline? [Partial Answer: about 350 kg lead-acid batteries]

c. Assume that 25% of the energy in the gasoline will be converted to useful mechanical energy by the engine in the car. Also assume that 90% of the electric energy in the battery system will be converted to useful mechanical energy by the electric motor in the car. (1) How many kg of lithium-ion batteries would it take to provide the same amount of useful mechanical energy as one kg of gasoline? (2) How many kg of lead-acid batteries would it take to provide the same amount of useful mechanical energy as one kg of gasoline? [Partial Answer: about 100 kg of lead-acid batteries]

2. **Figure 9.15** below is a rough illustration of what happens when discharging a lead-acid battery at different rates. Assume that the data is for a 12 V 60 amp-hour capacity lead-acid battery. The lowest curve on the graph is labeled 3.0 C, which means the discharge amperage was $3 \times 60 = 180$ amps. The second lowest curve labeled 2.0C means the battery is discharged at a rate of 120 amps. There are several other curves on the graph. Using the graph, estimate the amount of energy that can be drawn from the battery at the different discharge rates (3.0C, 2.0C, 1.0C, 0.4C, 0.2C, 0.05C, 0.01C).

    [Process: First use 5 to 7 points along the curve to find the average discharge voltage. Point 1 should be at one minute, i.e., do not use the first minute in finding the average voltage. Use the x-axis on the graph to get the number of hours of discharge and multiply the discharge amperage (180 amps for the lowest curve) by the hours to get the amp-hours discharge. The energy drawn from the battery is the average voltage multiplied by the amp-hours. Use Excel to plot the energy that can be drawn out of the battery vs the discharge amperage. [Partial Answer: At 1.0C there is about 480 W-h energy.]

FIGURE 9.15 Lead-acid battery discharge curves.

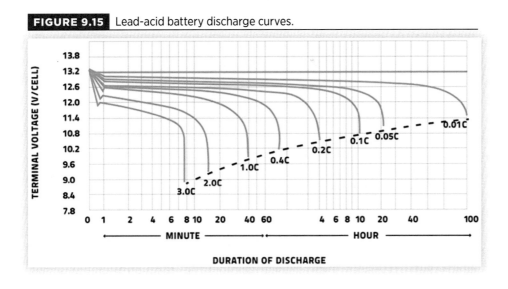

3. Data are collected for charging a small lithium-ion cell. For most of the charging cycle the battery is charged at a steady 1.3 amps. After 10 min of charging the voltage is observed to be 3.85 V. After 90 min of charging the voltage is observed to be 4.16 V. [Raising the charging voltage of a lithium-ion cell above 4.2 V will damage the battery, so after 90 min the charging amperage must decrease to avoid damaging the battery.]

   a. Assume that the battery voltage varies linearly between 10 min and 90 min. Calculate the energy put into the battery (watt-h) and the amp-hours of charge put into the battery during this 80 min of the charging cycle. Assume 100% charge efficiency.

   b. Assume that the amp-hours charge calculated above will be drawn out of the battery at an average of 3.7 V. Calculate the energy that will be drawn out of the battery. What is the energy efficiency of the battery? [Partial Answer: 92.4% efficient]

4. Suppose we have an 80-lb lead-acid battery. 10% of the weight of the battery is active material on the plates, i.e., we have a total of 8 lb of active Pb and PbO_2 on the plates that can be used to make electric power. Use the Gibbs free energy method to estimate the total stored energy in the battery. Assume that the open circuit voltage for a lead-acid cell is 2.1 V. [The answer should be a little less than 1 kilowatt-hour of energy.]

5. Assume that a large lead-acid for k lift battery is at a 50% state of charge. The battery is charged briefly at a rate of 50 amps and the voltage is observed to be 14.3 V. The battery is then discharged briefly at 50 amps and the voltage is observed to be 11.0 V. Use this data to estimate the internal resistance of the battery. [Hint: V0 + (50 amps) × R = 14.3 V]

   Calculate the energy efficiency for this battery if it is charged at 30 amps and discharged at 20 amps. [Answer: 87.9%]

6. Assume that a lead-acid cell is put on a trickle charger to maintain full charge on the battery. There is no real increase in charge in the battery. The energy going into the battery contributes to electrolysis of the water in the electrolyte. Assume the voltage across the cell is maintained at 2.2 V and one mole of water is converted to hydrogen and oxygen gas.

   a. How much energy in Joules is used to convert the water to hydrogen and oxygen gas?

   b. How much energy in watt-hours? [Answer: 118 W-h]

   c. How many grams of water in the electrolyte will be lost for each KWH of energy?

7. The goal for this homework problem is to have you think about the battery chemistry of lead-acid batteries. Lead-acid batteries will be important in small electric vehicles like forklifts and golf carts for a long time because they are inexpensive and reliable. Review the notes on lead-acid batteries and do a brief google search on lead-acid battery chemistry. Make a hand-drawn sketch of a lead-acid battery similar to the figure in the notes.

    a. Write a short paragraph explaining the chemistry on the positive plates.

    b. Write a short paragraph explaining the chemistry on the negative plates.

8. The most important battery technology in the near future is the lithium-ion battery. Lithium-ion batteries show the most promise for us in electric cars. Review the notes on lithium-ion batteries and do a brief google search on lithium-ion battery chemistry.

    a. Make a hand-drawn sketch of a lithium-ion battery similar to the figure in the notes.

    b. Write a short paragraph explaining the chemistry on the positive plates.

    c. Write a short paragraph explaining the chemistry on the negative plates.

    d. Write a short paragraph explaining how the electrolyte works.

9.6 Review of Unit 2

1. What is the average thermal efficiency of the equipment used for generating electricity in the USA? Show your calculations.

2. What is the average thermal efficiency of transportation in the USA? Show your calculations.

3. The Lawrence Livermore National Laboratory classifies a large part of the energy we use as "Rejected Energy". What does the term "Rejected Energy" mean?

4. The USA uses 97.5 Quads of energy annually. If there are 320 million people in the USA, how much energy on the average does each person use each day? Assuming 121,000 BTU per gallon for gasoline, express your answer in equivalent gallons of gasoline used each day.

5. Assuming that a horse can produce 4 KWH of energy each day, how many horses would it take to produce 97.5 Quads of energy each year?

6. Transportation produces 1830 million metric tons of carbon dioxide in the USA each year. How many moles of carbon dioxide gas does this represent?

7. The Lawrence Livermore National Laboratory regards biomass fuels such as alcohol and biodiesel to have zero carbon dioxide emissions. However, burning these fuels generates carbon dioxide. What is the reason for considering biomass fuels to have zero carbon dioxide emissions?

8. If we add transportation and generating electricity, what fraction of carbon dioxide generated in the USA comes from the sum of transportation and electricity generation?

9. Assume that worldwide we consume 170.5 Quads of petroleum, 146.6 Quads of coal, and 119.4 Quads of natural gas annually. On average, petroleum gives us 14 BTUs of energy per gram of CO_2 produced, coal gives us 9.5 BTUs energy per

gram of CO_2 produced, and natural gas gives us 19 BTUs of energy per gram of CO_2 produced. How much carbon dioxide is produced annually worldwide in millions of metric tons?

10. Propane has a chemical formula of C_3H_8. The energy density is 67,150 BTU/gallon, and the density is 1525 g/gal. Calculate the number of BTUs we get from propane for each gram of carbon dioxide generated. (Your answer should be about 14.7 BTU per gram CO_2.)

11. Assume that the atmosphere is 75.41% nitrogen (N_2), 23.20% oxygen (O_2) and 1.39% Argon (Ar) by weight. If the total mass of the atmosphere is 5.15×1018 kg, how many moles of nitrogen, oxygen and argon are in the atmosphere? Assume nitrogen atoms have an atomic weight of 14, oxygen 16, and argon 40. (Your answer should be 1.378×1020 moles nitrogen. Nitrogen is 78% of the molecules in the atmosphere.)

12. The cost of coal we are burning is such that we get about 400,000 BTUs of energy for each dollar we spend on coal. Assume that the coal plant is 33% efficient in converting the energy in the coal into electric power. What is the fuel cost per KWH of electric energy? (2.56 cents)

13. When purchasing natural gas, the power company gets 300,000 BTUs of energy for each dollar of natural gas purchased. Assume that the gas plant is 37% efficient in converting the energy in the natural gas to electric power. What is the fuel cost per KWH of electric energy? (3.04 cents)

14. A 65-m diameter wind turbine is operating in a 23-mph wind. The density of air is 1.22 kg/m³. Assume that the efficiency of the turbine is 47% and that the efficiency of the generator and inverter is 90%. What is the power output of the wind turbine in watts? (930,097 W)

15. Suppose the 65-m diameter wind turbine above has a capital cost of $2.5 million for purchase and installation. To justify the investment, we need to generate enough electricity annually for 7% of the investment capital, which is $175,000 annually. We estimate the wholesale value of the electricity to be 7 cents per KWH. Calculate the average wind speed required to justify the investment. (15.51 mph)

16. Calculate the angle of declination on March 24, 2017. If the city in question is located at a latitude of 33.72°, what is the angle between a normal to the surface of the earth and the sun at high noon? (1.411 declination angle, 32.31° at high noon)

17. On March 24, 2017 at 2:00 pm central daylight savings time, what is the longitudinal location of the sun? If the city in question is located at a longitude of −89.73°, what is the hour angle? (−105°, 15.27°)

18. Assume the declination angle is 32.31°, the hour angle is 15.27°, the latitude is 33.72° and the longitude is −89.73°. The solar array is tilted 18.4° to the south to keep it better aligned with the sun. Calculate the cosine of the angle between sun and the solar array.

19. Assume that producing 12.6 Quads of electric power also produces 2040 million metric tons of carbon dioxide. Calculate the grams of CO= produced per KWH of electric power. (1 KWH = 3412 BTU, 552.4 grams CO_2 per KWH)

20. Assume 552.4 g CO_2 are produced for each KWH of electric power. We have an electric car with a 60 KWH battery pack that is can 165 miles on the highway between charges. The batteries can be fast charged in two hours, but it will require 70 KWH of electric energy to charge them back to the 60 KWH of usable capacity. Calculate the carbon footprint of the electric vehicle in grams CO_2 per mile driven. (234.4 g CO_2 per mile)

21. Assume a pickup truck gets 25 mpg on the highway. Calculate the carbon footprint of the truck in grams CO_2 per mile driven. Assume that gasoline has 121,000 BTU/gallon and that we get 14.2 BTU of energy from gasoline for each gram of CO_2 produced. (340.8 grams CO_2 per mile)

Battery Review Problems

1. Estimate the capacity in W-h for a 1.0C discharge for the battery discharge curve shown in **Figure 9.16**. Use 5 points spaced along the curve. [Depends on points. About 7 W-h]

FIGURE 9.16 Lithium-ion discharge curves.

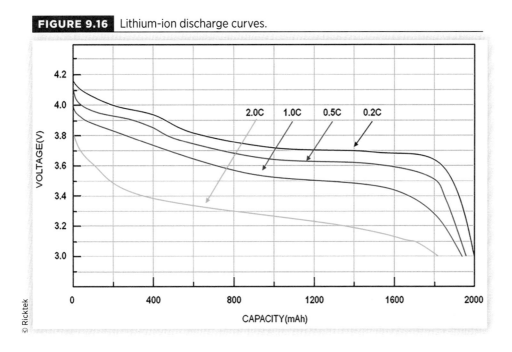

2. A deep cycle 12V lead-acid battery has a nominal capacity of 80 amp-h and weighs 72.8 lb.

    a. Calculate the energy density of the battery in MJ/kg. [2.2 lb = 1 kg] [0.1 MJ/kg]

    b. Suppose jet fuel has an energy density of 46.6 MJ/kg. The engine using the jet fuel is 35% efficient, but the electric motor using the battery power is 90% efficient. How many kg of batteries would be required to provide the same useful energy as from 1 kg of jet fuel? [About 173.5 batteries]

3. A battery is discharged at 1.2 amps and the average voltage during the discharge is 3.75 V. On a second test the battery is discharged at 1.5 amps and the average voltage during discharge is 3.69 V.

    a. Calculate the internal resistance of the battery. [0.2 ohms]

    b. Suppose the battery is charged at 0.8 amps and discharged at 0.9 amps. What is the energy efficiency of the battery under these charge/discharge conditions? [91.8%]

Unit 3

Detailed Modeling of Vehicle Efficiency

In Unit 1, we studied the energy consumption and fuel economy of vehicles at their cruising or operating speed. The energy consumed and fuel economy of the vehicle depends on the type of vehicle, the size and weight of the vehicle, and the thermal efficiency of the engine and drive system. When operating at the cruising or operating speed the following thermal efficiency ranges are typical for the internal combustion and jet engines in the different types of vehicles:

a. Cars and Light Trucks	25%–30%
b. Larger Trucks	30%–35%
c. Trains and Ships	30%–35%
d. Airplanes	25%–30%
e. Jets	35%

There is no reference I can give for the values above. After many years of modeling different types of vehicles for fuel economy, these are the thermal efficiency values that will give realistic estimates for the fuel economy and fuel consumption of the vehicle. The electric motors used in electric vehicles have a much higher thermal efficiency than the internal combustion engines used in most vehicles. The batteries required to store the electric energy are very heavy, and electric power is currently impractical for airplanes, jets, ships and large trucks. In many countries, trains are powered directly from the electric power grid and do not need to carry batteries. With the right infrastructure, trains can be electric powered using the technology available today. We are trying to develop electric powered cars and light trucks. The lithium-ion batteries are still too expensive and heavy for electric power to be practical for passenger cars and light trucks, but we are close enough to making it practical that we are continuing with development. We need to develop batteries that are less expensive and lighter in weight. Electric power is practical for smaller vehicles that do not need to travel long distances such as electric bicycles and scooters, wheelchairs and mobility vehicles, forklifts and vehicles used in manufacturing plants.

In Unit 2, we learned that the power plants that produce electricity are about 33% efficient overall in converting the primary energy sources into electricity. If we use fossil fuels to produce the electricity, then electric vehicles have a significant carbon emission because of the carbon dioxide produced in generating the electricity. In order to reduce our carbon emissions, we will need to convert vehicles to electric power and produce the electricity from sources other than fossil fuels.

Most vehicles spend their life operating at the cruising or operating speed. The methods taught in Unit 1 work well for estimating the energy consumption and fuel economy for large trucks, trains, ships, airplanes, and jets. Cars and light trucks spend a large percentage of their time in city traffic and the city driving has a large impact on their energy consumption and fuel economy. The thermal efficiency of the engine in city driving is lower than for highway driving. Electric powered and hybrid-electric powered cars and light trucks can be much more efficient in city driving than internal combustion powered cars and light trucks. In Unit 3, we will learn to do detailed modeling of cars and light trucks in city driving, suburban driving and highway driving and gain a better understanding of how it impacts the energy consumption and fuel economy of the vehicles.

10

Performance and Efficiency of Internal Combustion Engines

10.1. Engine Maps

Internal combustion engines operate over a range of revolutions per minute (RPMs) and torques. The engines can be tested, and a mapping of the engine performance as shown in **Figure 10.1** can be produced. The RPMs are on the x-axis and the torques are on the y-axis. The fuel consumption is measured in pound per horsepower-hour (lb/hp-h) in US customary units and in grams per kilowatt-hour (g/kW-h) in metric units. The highest thermal efficiency for this engine is the dark red region in the upper middle part of the graph. For the line encircling this region, the engine uses 0.411 lb of gasoline for each hp-h of energy produced by the engine. In the metric system, the engine uses 250 g of gasoline for each kW-h of energy produced.

The numbers on the graph can be used to obtain the thermal efficiency of the engine at different RPM and torque values. Gasoline has an energy density of 0.128 lb/hp-h, or in the metric system 77.85 g/kW-h. Dividing by the numbers on the graph gives the thermal efficiency of the engine. For example, dividing 0.128 lb/hp-h by 0.411 lb/hp-h yields a thermal efficiency of 31.1%. When operating the engine on the line bordering the red region, the engine is operating at a thermal efficiency of 31.1%. This means that 31.1% of the heat energy in the gasoline is being converted into useful mechanical energy and the other 68.9% is being converted into heat. In the metric system, divide 77.85 g/kW-h by 250 g/kW-h to get the 31.1% thermal efficiency of the engine. Let FC be the fuel consumption in lb/hp-h or g/kW-h.

$$\text{Thermal Efficiency} = \frac{0.128}{\text{FC}} \qquad \left(\text{with FC in lb / hp-h}\right) \qquad (10.1)$$

FIGURE 10.1 Engine map for Saturn 1.9L engine.

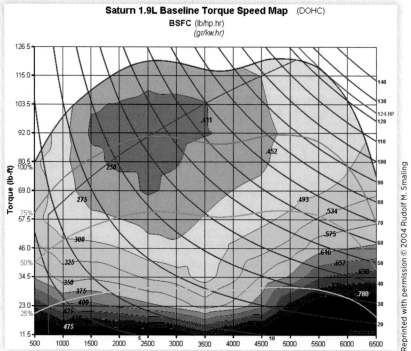

$$\text{Thermal Efficiency} = \frac{77.85}{FC} \quad \left(\text{with FC in g / kW-h}\right) \quad (10.2)$$

Example 10.1: Assume the engine is producing 23.0 ft-lb of torque at 1500 RPM and calculate the thermal efficiency of the engine.

Solution: From Figure 10.1, locate the point 1500 RPM and 23.0 ft-lb on the graph. The fuel consumption is between the 400 and 425 curves; a little closer to the 400 curve. Estimate the fuel consumption (FC) to be 410 g/kW-h when operating at 23.0 ft-lb torque and 1500 RPM and use Eq 10.2. Thermal efficiency is 77.85/410 = **19.0%**. In city, driving the engine will spend a lot of time operating at low torque and RPM and the thermal efficiency of the engine will be lower than when the vehicle is on the highway.

The blue lines in Figure 10.1 are lines of constant horsepower, with the numbers on the right side of the graph. For example, the engine can produce 20 hp by generating 34.5 ft-lb of torque at 3045 RPM, or it can generate 69.0 ft-lb of torque at 1522 RPM. Engine power is defined as the torque multiplied by the angular velocity. Assume the torque is to be in ft-lb, the angular velocity in RPM, and the power in hp. The following equation is a convenient way to relate the three quantities:

$$\text{Power}\left(\text{hp}\right) = \frac{\left[\text{Torque}\left(\text{ft-lb}\right)\right]\left(\text{RPM}\right)}{5252} \quad (10.3)$$

There is a pink line on the graph that shows how to get maximum efficiency out of the engine for the required power output. The pink line should be approximately perpendicular to the horsepower curves through the high-efficiency part of the mapping down to idle speed and then drops vertically. The transmission will be designed to try to follow this line, though it will not be able to exactly follow the line, especially for low power requirements. The transmission operates most efficiently in high gear, so overall efficiency will be a function of engine and transmission efficiency.

Three-speed transmissions were common for many years. As fuel economy and efficiency became more important more gears were added to the transmission. Having more gears allows the transmission access to the required power and have the engine operate at a higher thermal efficiency. Having the engine operate at a higher thermal efficiency will cause the vehicle to get better fuel economy. The newest cars have a continuous variable transmission (CVT). The electronics and software controlling the CVT will choose a gear ratio that allows the engine to operate at maximum thermal efficiency for the amount of horsepower it needs to deliver. For low power requirements, the CVT will push the engine to low RPMs to get the best thermal efficiency possible.

Example 10.2: Assume that the Saturn vehicle has a weight of 3000 lb, a rolling resistance coefficient of 0.010, a frontal area of 1.5 m², and a drag coefficient of 0.31. Assume level ground and the vehicle traveling at 40 mph. What is the ideal RPM for the engine to maximize the thermal efficiency of the engine? (Assume the transmission would be designed to shift and allow the engine to operate at maximum thermal efficiency for the required power.) Calculate the thermal efficiency of the engine using Figure 10.1. Assume that the transmission and final drive for the vehicle are 90% efficient in getting the engine power to the wheels and calculate the fuel economy of the vehicle in mpg.

Solution: When cruising at a steady speed on level pavement, the car must overcome rolling resistance and aerodynamic drag. From Unit 1 the power in hp is calculated as:

$$\text{Power}\,(\text{hp}) = \frac{(0.002667)\,W\,V\,Crr\left(1+\dfrac{V}{200}\right)+\left(7.184 \times 10^{-5}\right)V^3 A\,Cd}{\text{Transmission \& Drive Efficiency}} \qquad (10.4)$$

Plugging the given values into the equation yields:

$$\text{Power} = \frac{(0.002667)(3000)(40)(0.010)\left(1+\dfrac{40}{200}\right)+\left(7.184 \times 10^{-5}\right)40^3(1.5)(0.31)}{0.90} = 6.642\,\text{hp} \qquad (10.5)$$

The engine must provide 6.642 hp to push the car along at 40 mph on level pavement. According to Figure 10.1, at 6.642 hp the engine will operate most efficiently at 750 RPM, because that is where the 6.642 hp line would cross the pink line on the graph. The torque requirement is calculated as:

$$\text{Torque}\,(\text{ft-lb}) = \frac{(6.642\,\text{hp})(5252)}{750\,\text{RPM}} = 46.5\,\text{ft-lb} \qquad (10.6)$$

The point for 750 RPM and 46.51 ft-lb is in the green region in Figure 10.1 between the 300 and 325 lines, but closer to the 325 line. Estimate the value to be 320 g/kW-h.

$$\text{Thermal Efficiency} = \frac{77.85}{320} = 24.3\% \tag{10.7}$$

This is the thermal efficiency of the engine when delivering 6.642 hp at 750 RPM and 46.51 ft-lb torque. Gasohol will have an energy density of about 121,000 BTU/gal, and 23.4% of the energy in the fuel will be converted to useful mechanical energy to power the car. The rate of fuel consumption, or fuel power is calculated as:

$$\text{Fuel Power} = \frac{6.642\,\text{hp}}{0.243} = 27.33\,\text{hp} \tag{10.8}$$

For unit conversions, 1 hp is equal to 2544.4 BTU/hr. Putting this together:

$$(27.33\,\text{hp})\left(\frac{2544.4\,\text{BTU/h}}{\text{hp}}\right)\left(\frac{\text{gal}}{121{,}000\,\text{BTU}}\right) = 0.575\,\text{gph} \tag{10.9}$$

Driving on level ground at 40 mph the car would use 0.575 gallons of gas each hour. The fuel economy for the car is the speed divided by the gph fuel consumption.

$$\text{Fuel Economy} = \frac{40\,\text{mph}}{0.575\,\text{gph}} = 69.6\,\text{mpg} \tag{10.10}$$

This seems like a very high fuel economy for the car. The Environmental Protection Agency (EPA) rating is less than 69.6 mpg for city or highway driving. But if the car were operated at 750 RPM and 40 mph steady speed it would get close to the 69.6 mpg calculated. As we move forward with the discussion, I will explain why the transmission will not actually choose to have the engine operate at 750 RPM. It would make the car unpleasant to drive. The transmission will have the engine operate at a higher RPM, which reduces the thermal efficiency of the engine, which leads to a more realistic estimate of the mpg fuel economy.

Example 10.3: Operating the car at 750 RPM will make the car feel sluggish if you want to accelerate or if you start up a slight incline in the road. If you look at Figure 10.1, the maximum possible horsepower for the engine at 750 RPM is about 10 hp. When you mash the foot pedal you want the engine to produce more power so you can speed up and the engine cannot produce much more power at 750 RPM. You will feel like you are driving a very underpowered vehicle and you will hate the car. The car would be very fuel-efficient, but Saturn would not sell many because the car is unpleasant to drive.

Solution: To help alleviate the sluggishness let's assume that the transmission causes the engine to operate at 1500 RPM and calculate the fuel economy for steady cruising at 40 mph on level pavement. At 1500 RPM, the engine could produce a little over 30 hp, and that will make the car accelerate better and be more pleasant to drive.

The power the engine must produce to push the car along at 40 mph is the same 6.642 hp regardless of the engine RPM. The torque for 1500 RPM is calculated as:

$$\text{Torque (ft-lb)} = \frac{(6.642\,\text{hp})(5252)}{1500\,\text{RPM}} = 23.26\,\text{ft-lb} \qquad (10.11)$$

Locating the point 1500 RPM and 23.26 ft-lb torque in Figure 10.1 puts us in the lavender region between the 400 and 425 lines, but closer to the 400 line. Assume that the fuel consumption is 410 g/kW-h. The thermal efficiency is calculated as:

$$\text{Thermal Efficiency} = \frac{77.85}{410} = 19.0\% \qquad (10.12)$$

Notice that the thermal efficiency of the engine is reduced from 24.3% at 750 RPM to 19.0% at 1500 RPM. This increases the amount of fuel that must be consumed to generate the 6.642 hp required to push the car along at 40 mph.

$$\left(\frac{6.642\,\text{hp}}{0.190}\right)\left(\frac{2544.4\,\text{BTU/h}}{\text{hp}}\right)\left(\frac{\text{gal}}{121{,}000\,\text{BTU}}\right) = 0.735\,\text{gph} \qquad (10.13)$$

$$\text{Fuel Economy} = \frac{40\,\text{mph}}{0.735\,\text{gph}} = 54.4\,\text{mpg} \qquad (10.14)$$

This is still a good fuel economy and the car will be much more pleasant to drive if it is geared like this. If you were to take the actual production car and put it on a level track driving at 40 mpg it would deliver a fuel economy near the 54.4 mpg calculated.

Example 10.4: Rework the problem for highway mileage assuming the car is traveling at 70 mph. What is the ideal RPM for the engine and what is the fuel economy?

Solution: The power required to push the car along at 70 mph on the level pavement is calculated as:

$$\text{Power} = \frac{(0.002667)(3000)(70)(0.010)\left(1+\frac{70}{200}\right) + (7.184 \times 10^{-5})70^3(1.5)(0.31)}{0.90} = 21.13\,\text{hp} \qquad (10.15)$$

The 21.13 hp curve crosses the pink line on the graph at about 1550 RPM. The torque required to produce 22.24 hp at 1550 RPM is 71.6 ft-lb. This falls in the orange region on the graph between the 250 and 275 lines, but closer to the 250 line. Assume the fuel consumption is 260 g/kW-h. This yields a thermal efficiency of 77.85/260 = 29.9%, fuel consumption of 1.484 gph, and a fuel economy of 47.2 mpg. If the transmission were geared so that the engine operated at 1550 RPM at 70 mph, the car would get approximately 47.2 mpg fuel economy. But this will make the car feel a little sluggish on the highway if you want to pass someone. To make the car more pleasant to drive, the transmission will be geared to a higher RPM on the highway.

Example 10.5: To balance drivability with fuel economy, the manufacturer (Saturn) chose to have the engine turn 2500 RPM at 70 mph. Calculate the fuel economy for the car traveling at 70 mph and the engine turning 2500 RPM.

Solution: The power required is 21.13 hp for driving at 70 mph. If the engine is operating at 2500 RPM, Eq. 10.3 is used to calculate the torque to be 44.39 ft-lb. This is in the yellow region of the graph approximately halfway between 275 and 300 lines. Assume a fuel consumption of 287 g/kW-h. The thermal efficiency is 77.85/287 = 27.12%. Fuel consumption is 1.638 gph and fuel economy is 42.7 mpg.

The EPA fuel economy estimate for this vehicle was 40 mpg for highway driving. The purpose of the four examples was to illustrate how to use the engine map (Figure 10.1) to calculate the fuel economy of the vehicle. In city driving the car will accelerate and decelerate, and the engine will operate over a range of RPMs and torques. To calculate fuel economy for city driving, we have to calculate the instantaneous fuel economy and then numerically integrate over a typical city driving pattern. The approach illustrated in the four examples above will be used later in the unit to calculate the fuel economy for city driving. Before we get to the detailed modeling there are a few more important topics to be discussed.

The size of the engine is an important factor in the fuel economy of the vehicle. In general, putting a larger engine in a vehicle will yield a lower fuel economy. It is important to understand why this is true. Steady speed driving, even at highway speeds, doesn't use enough horsepower for us to get into the highest efficiency range for the vehicle. If the engine were used to power a generator, we could gear it to operate at the highest efficiency, but for a car, this is not possible.

Example 10.6: What if we use a smaller engine to power the car, so we can operate at a higher efficiency when driving on the highway?

Solution: The short answer is that the car will get better fuel economy on the highway because the engine will operate at a higher thermal efficiency. Car performance would suffer, and people would not enjoy driving the car with the smaller engine, but the fuel economy would improve.

Examining the engine map (Figure 10.1) for the Saturn car the maximum thermal efficiency is in the middle of the red range at about 2500 RPM and 92.0 ft-lb of torque. The maximum horsepower the engine can produce is 124 hp, but at maximum thermal efficiency, the engine is producing 43.8 hp.

$$\frac{(2500\,\text{RPM})(92.0\,\text{ft} - \text{lb})}{5252} = 43.8\,\text{hp} \tag{10.16}$$

From previous examples, we calculated that it requires 21.13 hp to push the car along on level pavement at 70 mph. If we were to cut the engine size in half, the maximum rated horsepower would be 62 hp, and the engine would deliver maximum thermal efficiency at half of 43.8 hp, which is 21.4 hp. The conclusion is that using an engine half the size of the one Saturn put in the car would allow the car to operate at maximum thermal efficiency when driving on the highway. The smaller engine would operate at 2500 RPM and produce 44.39 ft-lb of torque. The fuel consumption when operating at that torque and RPM would be approximately 230 g/KWH. The thermal efficiency of the engine would be 77.85/230 = 33.85%. Fuel consumption and economy are calculated as follows:

$$\left(\frac{21.13\,\text{hp}}{0.190}\right)\left(\frac{2544.4\,\text{BTU/h}}{\text{hp}}\right)\left(\frac{\text{gal}}{121{,}000\,\text{BTU}}\right) = 1.313\,\text{gph} \qquad (10.17)$$

$$\text{Fuel Economy} = \frac{70\,\text{mph}}{1.313\,\text{gph}} = 53.3\,\text{mpg} \qquad (10.18)$$

Using an engine half the size that Saturn actually used would raise the fuel economy from 42.7 mpg to 53.3 mpg, an approximate 25% improvement in fuel economy. This is a large improvement in fuel economy but driving a car with an engine half the size will not be much fun to drive. Part of the reason Saturn chose to put a 124 hp engine in the car instead of a 62 hp engine is to provide adequate acceleration of the vehicle.

Acceleration of the car is important. The engine must have enough horsepower to accelerate the car properly around town and on the highway. Testing indicates that around town people expect to be able to accelerate at about 7 ft/s². Sports cars will be able to accelerate faster than this, but for most people, an acceleration of 7 ft/s² will feel solid and allow the car to keep up in traffic. Car acceleration is commonly measured as the time it takes to accelerate from zero to 60 mph. Since 60 mph is equal to 88 ft/s, the time for a 7 ft/s² acceleration is calculated as:

$$0-60\,\text{mph time} = \frac{88\,\text{ft/s}}{7\,\text{ft/s}^2} = 12.6\,\text{s} \qquad (10.19)$$

The fastest sports cars will accelerate zero to 60 mph in about half this amount of time, which means they can provide 14 ft/s² acceleration. If you were to mash the pedal to the floor, the Saturn would be able to accelerate faster than 7 ft/s², but for normal driving, people do not mash the pedal to the floor. A 7 ft/s² acceleration is acceptable for normal city driving. We sometimes measure acceleration in g's, and the acceleration of gravity is 32.2 ft/s². Another way to look at the acceleration is:

$$\text{Acceleration} = \frac{7\,\text{ft/s}^2}{32.2\,\text{ft/s}^2} = 0.22\,\text{g acceleration} \qquad (10.20)$$

The engine must be able to provide the horsepower required for acceleration. The weight of the Saturn car we have been discussing is 3000 lb. The mass of the car is 3000 lb/32.2 ft/s² = 93.2 slug. The force required to accelerate at 7 ft/s² is mass multiplied by acceleration:

$$\text{Force} = (93.2\,\text{slug})(7\,\text{ft/s}^2) = 652.2\,\text{lb} \qquad (10.21)$$

The engine power required for acceleration is force multiplied by speed. One hp is 550 ft-lb/s, so the power requirements at 15 mph, 30 mph, and 60 mph can be calculated as:

a. 15 mph = 22 ft/s, Power = (652.2)(22)/(550) = 26.1 hp

b. 30 mph = 44 ft/s, Power = (652.2)(44)/(550) = 52.2 hp

c. 60 mph = 88 ft/s, Power = (652.2)(88)/(550) = 104.3 hp

The calculations illustrate that the engine horsepower must increase as speed increases to provide the same acceleration. Accelerating the car at highway speed requires a lot more horsepower than accelerating in city traffic. Most cars do not have enough horsepower to accelerate at 7 ft/s² when traveling at highway speed. Drivers do not expect the car to be able to accelerate at 7 ft/s² on the highway. An acceleration of 4 ft/s² is acceptable on the highway.

The calculations above are the horsepower required for acceleration. The power to overcome rolling and aerodynamic drag must be added and the efficiency of the drive system must also be factored in. The engine will need to produce more horsepower than the numbers above. For a solid acceleration like 7 ft/s², most of the engine horsepower goes to provide the acceleration, but a significant percentage goes for rolling and aerodynamic drag and inefficiencies in the drive system. The hp calculations in a, b, and c above are reasonable, but low estimates of the engine horsepower are required.

Around town, the 62 hp engine will have enough horsepower to accelerate the car up to approximately 25 mph, and then the car will feel sluggish and unable to keep up in traffic. The driver will mash the pedal to the floor all the time trying to keep up in traffic. After a test drive, not many people will be willing to buy the car. The 124 hp engine that Saturn selected is adequate to provide the acceleration and allow the car to keep up in traffic. It's not a race car, but it is adequate.

If the car were fitted with a larger engine, the acceleration would improve, but acceleration is also limited by the friction between the tires and the pavement. If the engine is powerful enough to spin the tires, then acceleration is limited by friction rather than by engine horsepower. As the car accelerates, the weight shifts to the rear tires to some extent. The friction force is equal to the weight on the tires multiplied by the coefficient of friction. Since there is more weight on the rear tires, rear-wheel drive cars can accelerate faster than front-wheel drive cars. All the really fast cars are rear-wheel drive, including NASCAR, Indy Car, Dragsters, Corvette, Camaro, Mustang, Charger, Porsche, Ferrari, etc. When designing for high acceleration, rear-wheel drive is the best choice.

In theory, four-wheel drive cars should be able to provide more acceleration than rear-wheel drive cars, but there are control issues associated with dividing the power between the front and rear wheels and not causing the front wheels to slip. It is a difficult problem to solve. From a practical standpoint, all the fastest cars have been rear-wheel drive. The focus of this book is on energy efficiency, but at this point, it is important to do the dynamic analysis associated with the acceleration of a car.

Example 10.7: Assume a 3000 lb car with the center of gravity (CG) 2 ft above the pavement and centered between the front and rear axle as illustrated in **Figure 10.2**. Assume that the car is to accelerate at 13 ft/s². Find the reactions on the front and rear (N_F and N_R) and the traction force T required to provide the acceleration.

$$\sum F_x = 0 = 1211\,\text{lb} - T \tag{10.22}$$

$$\sum F_y = 0 = N_F + N_R - 3000\,\text{lb} \tag{10.23}$$

$$\sum M_{N_F} = 0 = N_R(10) - 3000(5) - 1211(2) \tag{10.24}$$

FIGURE 10.2 Dynamics of acceleration for Example 10.7

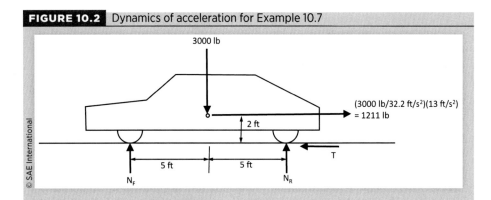

Solution: Solving the equations the traction force T = 1211 lb, N_R = 1742 lb, and N_F = 1258 lb. Front-wheel drive cars have the traction force on the front tires and rear-wheel drive cars have the traction force on the rear wheels. The minimum coefficient of friction required to prevent slipping is equal to the traction force divided by the normal force.

$$\text{Front Drive Minimum Coefficient of Friction} = \frac{1211}{1258} = 0.963 \quad (10.25)$$

$$\text{Rear Drive Minimum Coefficient of Friction} = \frac{1211}{1742} = 0.695 \quad (10.26)$$

A conservative estimate for the coefficient of friction between tires and the dry pavement is 0.80. From the calculations above, the rear drive car would be able to achieve an acceleration of 13 ft/s² without slipping the tires because the required coefficient of friction is 0.695 and the actual coefficient of friction is 0.80. A front-wheel drive car would not be able to achieve an acceleration of 13 ft/s² because the required coefficient of friction is 0.963 and the actual coefficient of friction is 0.80. The front-wheel drive car can only achieve an acceleration of about 11 ft/s² before the front tires begin to slip. The rear-wheel drive car can achieve an acceleration of about 15 ft/s² before the rear wheels begin to slip.

The problem above can be set up in more general terms as illustrated in **Figure 10.3**:

$$\text{Front Drive Maximum Acceleration} = \frac{\mu b g}{(a + b + \mu c)} \quad (10.27)$$

$$\text{Rear Drive Maximum Acceleration} = \frac{\mu b g}{(a + b - \mu c)} \quad (10.28)$$

Where μ is the coefficient between the tires and pavement, which is typically 0.80, and g is the acceleration of gravity. If the values a = b = 5 ft, c = 2 ft are substituted into the equations above the maximum acceleration for front drive cars is 11.1 ft/s² and for rear drive cars it is 15.3 ft/s². Eq. 10.27 and 10.28 form a nice solution to the problem,

FIGURE 10.3 Dynamics of acceleration, general case.

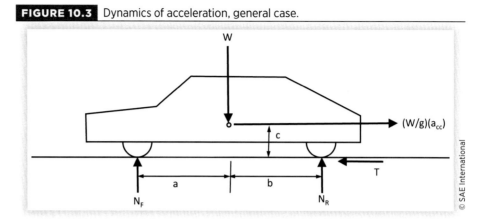

© SAE International

but it is very difficult to get accurate values for the a, b, and c distances on production cars, and the 0.80 coefficient of friction between the tires and the pavement depends on the tires and the pavement. Sports cars will use tires that have a higher coefficient of friction than the tires used on typical sedans.

Since many people are familiar with the zero to 60 mph acceleration times it seems prudent at this point to calculate the zero to 60 mph times for front-wheel drive and rear-wheel drive cars using the values above (60 mph = 88 ft/s).

$$\text{Front Drive:} \quad \frac{88 \, \text{ft/s}}{11.1 \, \text{ft/s}^2} = 7.93 \text{s} \tag{10.29}$$

$$\text{Rear Drive:} \quad \frac{88 \, \text{ft/s}}{15.3 \, \text{ft/s}^2} = 5.75 \text{s} \tag{10.30}$$

There are very few production cars that can accelerate from zero to 60 mph in 5.75 s. There are a few expensive sports cars that can accelerate zero to 60 mph in less than 5.75 s because they have very powerful engines and tires that provide excellent traction. There are a few expensive front drive cars that can come close to the zero to 60 mph time of 7.93 s. Most rear drive cars can accelerate at 15 ft/s² at low speed and most front drive cars can accelerate at 11 ft/s² at low speeds. As speed increases the horsepower required to sustain the acceleration increases and most production cars do not have enough horsepower to sustain the acceleration up to 60 mph. The relatively inexpensive 4-cylinder cars and small sport utility vehicles (SUVs) that many people drive can accelerate from zero to 60 mph in about 12 s. That is an acceptable acceleration for most people.

To model acceleration of a car, we need to consider that the acceleration is limited by the traction between the tires and the pavement and by the horsepower of the engine. For lower speeds, the engine will have enough horsepower to slip the tires and traction will be what limits the acceleration of the car. For this unit, we will assume front-wheel drive cars are limited to 11 ft/s² acceleration and rear-wheel drive cars are limited to 15 ft/s² acceleration. Actual acceleration limits vary from car-to-car as illustrated in Figure 10.3 and Eqs. 10.27 and 10.28. Using limits of 11 ft/s² and 15 ft/s² will be close to the correct values and will allow us to talk about the differences in front and rear drive cars.

As the car speeds up, there comes a point where the engine cannot provide enough horsepower to slip the tires, and at that point, it is the horsepower of the engine that limits the acceleration of the car. From a practical standpoint, most people will not push the engine to the maximum possible horsepower. For example, the Saturn engine is rated at 124 hp, but to get 124 hp from the engine the drive would need to have the engine turning 5700 RPM, and most people are not going to push the engine that hard. At 4500 RPM, the engine will produce about 90 hp, which is about 75% of the rated horsepower. From a practical viewpoint, most people will never have the engine producing more than 75% of the rated horsepower.

We will use Excel to develop a model for the acceleration of the car. In the model there will be a logic statement such that:

a. Acceleration cannot be more than 11 ft/s^2 for front-wheel drive or 15 ft/s^2 for rear-wheel drive cars. This is the traction limit for acceleration.

b. Acceleration cannot be more than $(0.75\,P_{max})/(mass*speed)$. P_{max} is the maximum rated horsepower for the engine, and the driver will not use more than 75% of the rated horsepower. This is the power limitation for acceleration.

Example 10.8: Model acceleration of the Saturn car starting from rest for 20 s of acceleration. Assume the weight is W = 3000 lb, rolling coefficient is Crr = 0.010, drag area is A = 1.5 m^2, drag coefficient is Cd = 0.31, maximum rated engine power is P_{max} = 62 hp, maximum traction acceleration is 11 ft/s^2 (front-wheel drive), transmission/ drive efficiency = 0.88.

Solution: In a previous example we discovered that using a small 62 hp engine in the Saturn car would allow it to achieve about 53.3 mpg fuel economy on the highway, which is significantly better than the 42.7 mpg highway fuel economy the car achieves with the 124 hp engine that came in the car. The purpose of this example is to illustrate quantitatively the sluggish performance with the small engine and help you understand why we would not want to put such a small engine in the car. Table 10.1 illustrates the acceleration spreadsheet model for the Saturn car for the first 2.6 s.

In the first row, we start with time and speed equal to zero. Since the speed is zero, the rolling and aero power are both zero too. The maximum power that can be delivered to the wheels is 75% of the maximum rated engine power multiplied by the transmission/ drive efficiency. The equation for maximum power acceleration has the speed term in the denominator, so for zero speed, the equation yields infinity. The logic statement makes the maximum acceleration equal to the traction acceleration, which in this case is 11 ft/s^2.

a. In this model, the time step was chosen as 0.1 s. The column for the time was incremented at 0.1 s to 20 s to model the first 20 s of acceleration.

b. In the second column for speed, the speed is equal to the initial speed at the beginning of the time step plus the acceleration multiplied by the time step. For the second row the speed is 0 ft/s plus (11 ft/s^2)(0.1 s) = 1.1 ft/s. For the third row the speed is 1.1 ft/s plus (11)(0.1) = 2.2 ft/s.

c. In the third column, the speed is converted to mph because we think of speed in terms of mph and it will make more sense when we graph it. We also need the speed in mph to be able to use the equations for rolling and aero drag. To convert speed in ft/s to mph we multiply by (3600/5280).

d. In the fourth column, the rolling and aero drag are calculated using equations from Unit 1. Rolling + Aero = $(0.002667)WVCrr(1 + V/200) + (7.184 \times 10^{-5})V^3ACd$. Rolling + Aero is small and insignificant at low speed as illustrated from the numbers in the table, but as the car speeds up it becomes more significant.

e. In the fifth column, the available power for acceleration is the 40.92 hp the engine can deliver to the wheels minus the power to overcome rolling and aero drag.

f. In the sixth column, the maximum power acceleration is calculated as: Maximum Power Acceleration = (available power)(550)/(mass*speed). The 550 term converts horsepower to ft-lb/s and the speed, in this case, should be in ft/s. At low speeds, the engine has plenty of power to spin the wheels, but as the car speeds up the maximum power acceleration will become less than 11 ft/s² and will be what limits the acceleration of the car.

g. The 7th column is a logic statement that chooses the smaller value of the maximum power acceleration in column 6 and the maximum traction acceleration of 11 ft/s². The acceleration will be 11 ft/s² until the car reaches a speed where the engine no longer has enough horsepower to continue accelerating at 11 ft/s², and then the acceleration will decrease. For this example, the transition from traction limited acceleration to hp limited acceleration happens at 2.0 s.

Once the first few lines of the Excel program are developed it is possible to select the third line and copy it to the bottom of the spreadsheet at time = 20 s. There are details of indexing in Excel, and it is assumed that the reader has a good understanding of how to use Excel. The results for the 62 hp engine are illustrated in the figure below.

The graph in **Figure 10.4** shows that it will take 20 s for the car to accelerate to 60 mph, i.e., the zero to 60 mph time is 20 s. This will seem very sluggish and underpowered

FIGURE 10.4 Acceleration with 62 hp engine.

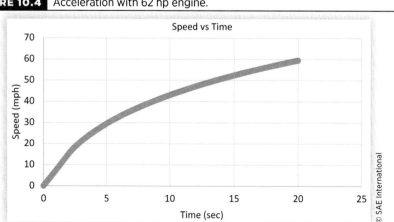

© SAE International

compared to the cars we drive. Looking at the numbers in the spreadsheet at a time of 2 s the engine no longer has enough horsepower to accelerate at 11 ft/s². As the car speeds up the power to overcome rolling and aero drag increases, which reduces the power available for acceleration and the acceleration of the car. The maximum top speed for this vehicle would be when the rolling and aero power total to the 40.92 hp the engine can provide to the wheels. In this case, the maximum top speed is 96 mph, and it takes a long time to get to that speed.

The car comes with a 124 hp engine, and if the parameters in the spreadsheet model are changed, the car is able to accelerate to 60 mph in 10.7 s, which is typical to the cars and trucks that we drive. If the spreadsheet was developed correctly it is only necessary to change the rated power to 124 hp and the results should be recalculated. The 124 hp engine has enough hp to accelerate the car at 11 ft/s² for 3.8 s, and then acceleration is limited by engine hp. The results are illustrated in **Figure 10.5**.

FIGURE 10.5   Acceleration with 124 hp engine.

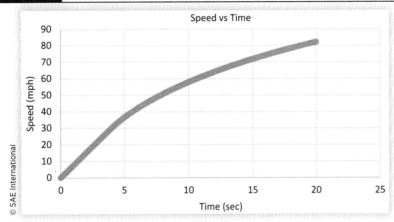

Hill Climbing: Another aspect of normal driving is that the car will need to have adequate power to drive uphill. For hill climbing, the power to overcome gravity is equal to the gravitational force pushing the car downhill multiplied by the vertical speed component of the car (**Figure 10.6**).

If the car has a weight W and is traveling at speed V up a hill of angled slope θ, the power to overcome gravity is:

$$\text{Gravitational Power} = W\,V\sin(\theta) \qquad (10.31)$$

FIGURE 10.6   Car traveling uphill.

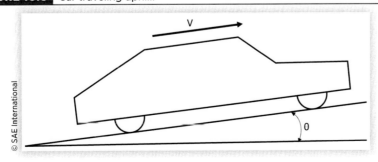

For highways, we normally measure the slope in grade such that the grade is the sine of the angle. For example, a 6% grade means the sine of the angle θ is 0.06. Most interstate highways limit the grade to 4%, though in mountainous areas the grades may be as high as 7%. Local roads may have grades as high as 11%, though it is rare. An 11% grade is very steep.

Example 10.9: Calculate the horsepower required to overcome gravity for a 5000-lb car traveling 70 mph up a 4% grade (70 mph = 102.67 ft/s).

Solution: Use Eq. 10.31.

$$\text{Gravitational Power} = \frac{(5000\,\text{lb})(102.67\,\text{ft-lb/s})(.04)}{550\,\dfrac{\text{ft-lb/s}}{\text{hp}}} = 37.33\,\text{hp} \qquad (10.32)$$

Example 10.10: Assume that the Saturn car is accelerating uphill. Modify the spreadsheet (Table 10.1) to account for the gravity loading in addition to the aerodynamic and rolling power.

Solution: The spreadsheet was set up so that the grade of the hill could be adjusted. An additional column was added for the hill-climbing power, which was subtracted to reduce the available power for acceleration. A 4% grade was used for the example below with the small 62 hp engine (Table 10.2).

Figure 10.7 is a graph showing the speed of the Saturn car for accelerating up a 4% grade for the first 20 sections. Notice that it takes 20 s to accelerate to 50 mph. It would take a very long time for the car to accelerate to highway speed if the ramp for entering the highway had a 4% uphill grade.

Figure 10.8 is the same graph as Figure 10.7 except the engine size was changed to 124 hp. With the larger engine, the car could accelerate upto a 4% grade to highway speed in less than 20 s. This is much more reasonable performance for the car.

The examples used for the Saturn car comparing the 62 hp engine and the 124 hp engine that comes in the car are meant to help the reader understand why the small engine is inadequate. The smaller engine will deliver better fuel economy, but it will not be able to provide the acceleration and hill-climbing capability we expect from our cars and trucks. Very few people would buy such an underpowered car. The manufacturer must balance fuel economy with performance. We all know that in the abstract. The purpose of this unit is to teach you how to do the calculations and understand them quantitatively.

The next step is to look at the engine map for a typical diesel engine. The mapping below is for a Volkswagen 2.0 TDI diesel engine. The colored curves are of constant horsepower and the numbers are in grams fuel per kW-h energy from the engine. Diesel

TABLE 10.1 Acceleration of Saturn car.

Weight	3000	lb		Rated Power	62	hp
Crr	0.01			Max Acc	11	ft/s^2
A	1.5	m^2		75% Power	46.5	hp
Cd	0.31			trans eff	0.88	
sec	ft/s	mph	Rolling & Aero	Available Acceleration	ft/s^2	ft/s^2
Time	Speed	speed	Power (hp)	Power (hp)	Max Acc	Cleaned
0	0.000	0.000	0.000	40.920	Infinity	11.000
0.1	1.100	0.750	0.060	40.860	219.281	11.000
0.2	2.200	1.500	0.121	40.799	109.477	11.000
0.3	3.300	2.250	0.182	40.738	72.875	11.000
0.4	4.400	3.000	0.245	40.675	54.573	11.000
0.5	5.500	3.750	0.307	40.613	43.591	11.000
0.6	6.600	4.500	0.371	40.549	36.269	11.000
0.7	7.700	5.250	0.436	40.484	31.038	11.000
0.8	8.800	6.000	0.502	40.418	27.114	11.000
0.9	9.900	6.750	0.569	40.351	24.061	11.000
1.0	11.000	7.500	0.637	40.283	21.619	11.000
1.1	12.100	8.250	0.706	40.214	19.620	11.000
1.2	13.200	9.000	0.777	40.143	17.953	11.000
1.3	14.300	9.750	0.849	40.071	16.542	11.000
1.4	15.400	10.500	0.923	39.997	15.332	11.000
1.5	16.500	11.250	0.998	39.922	14.283	11.000
1.6	17.600	12.000	1.075	39.845	13.365	11.000
1.7	18.700	12.750	1.154	39.766	12.553	11.000
1.8	19.800	13.500	1.235	39.685	11.832	11.000
1.9	20.900	14.250	1.318	39.602	11.186	11.000
2	22.000	15.000	1.403	39.517	10.604	10.604
2.1	23.060	15.723	1.487	39.433	10.095	10.095
2.2	24.070	16.411	1.568	39.352	9.651	9.651
2.3	25.035	17.069	1.648	39.272	9.260	9.260
2.4	25.961	17.701	1.727	39.193	8.912	8.912
2.5	26.852	18.308	1.804	39.116	8.599	8.599
2.6	27.712	18.895	1.880	39.040	8.316	8.316

TABLE 10.2 Acceleration of Saturn car uphill.

Weight	3000	lb								Rated Power	62	hp
Crr	0.01									Max Acc	11	ft/s^2
A	1.5	m^2			Grade					75% Power	46.5	hp
Cd	0.31				4	%				trans eff	0.88	

				Rolling & Aero	Hill Climbing	Available Acceleration		
sec	ft/s		mph	Power (hp)	Power (hp)	Power (hp)	ft/s^2	ft/s^2
Time	Speed		speed				Max Acc	Cleaned
0	0.000		0.000	0.000	0.000	40.920	Infinite	11.000
0.1	1.100		0.750	0.060	0.240	40.620	217.993	11.000
0.2	2.200		1.500	0.121	0.480	40.319	108.189	11.000
0.3	3.300		2.250	0.182	0.720	40.018	71.587	11.000
0.4	4.400		3.000	0.245	0.960	39.715	53.285	11.000
0.5	5.500		3.750	0.307	1.200	39.413	42.303	11.000
0.6	6.600		4.500	0.371	1.440	39.109	34.981	11.000
0.7	7.700		5.250	0.436	1.680	38.804	29.750	11.000
0.8	8.800		6.000	0.502	1.920	38.498	25.826	11.000
0.9	9.900		6.750	0.569	2.160	38.191	22.773	11.000
1.0	11.000		7.500	0.637	2.400	37.883	20.331	11.000
1.1	12.100		8.250	0.706	2.640	37.574	18.332	11.000
1.2	13.200		9.000	0.777	2.880	37.263	16.665	11.000
1.3	14.300		9.750	0.849	3.120	36.951	15.254	11.000
1.4	15.400		10.500	0.923	3.360	36.637	14.044	11.000
1.5	16.500		11.250	0.998	3.600	36.322	12.995	11.000
1.6	17.600		12.000	1.075	3.840	36.005	12.077	11.000
1.7	18.700		12.750	1.154	4.080	35.686	11.265	11.000
1.8	19.800		13.500	1.235	4.320	35.365	10.544	10.544
1.9	20.854		14.219	1.315	4.550	35.055	9.923	9.923
2	21.847		14.895	1.391	4.767	34.762	9.393	9.393
2.1	22.786		15.536	1.465	4.972	34.484	8.934	8.934
2.2	23.679		16.145	1.537	5.166	34.217	8.530	8.530
2.3	24.532		16.727	1.607	5.353	33.961	8.172	8.172
2.4	25.350		17.284	1.675	5.531	33.714	7.851	7.851
2.5	26.135		17.819	1.742	5.702	33.476	7.562	7.562
2.6	26.891		18.335	1.807	5.867	33.246	7.298	7.298

FIGURE 10.7 Uphill acceleration with 62 hp engine.

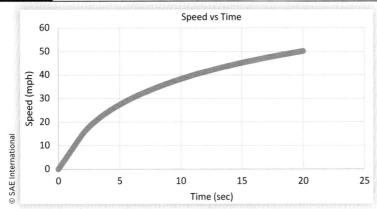

FIGURE 10.8 Uphill acceleration with 124 hp engine.

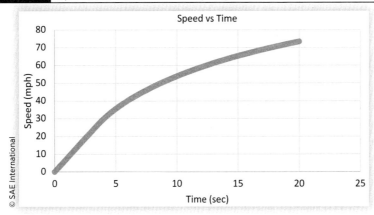

fuel has an energy density of 69.5 g/kW-h. The maximum efficiency is at about 2250 RPM and 84 hp, where the engine is using 196 g/kW-h fuel consumption. The maximum thermal efficiency of the engine is calculated as 69.5/196 = 35.5% thermal efficiency.

The y-axis in the engine map (**Figure 10.9**) is given in Brake Mean Effective Pressure (BMEP). Plotting BMEP on the y-axis normalizes the mapping with respect to the size of the engine. When torque is plotted on the y-axis large engines will have much higher torque values than small engines, and it is harder to compare the efficiency of large and small engines. BMEP will be the same for large and small engines and makes it easier in some ways to compare large and small engines. Many engineers prefer BMEP plotted on the y-axis rather than torque.

BMEP can be converted to torque values if you know the displacement volume of the engine (DV) and whether it is a 2-cycle or 4-cycle engine. All cars and trucks use 4-cycle engines, so Eq. 10.33 below is developed assuming a 4-cycle engine. The 4π factor in the equation is 2π for 2-cycle engines.

$$\text{Engine Torque} = \frac{(DV)(BMEP)}{4\pi} \ \left(\text{assuming 4 cycle engine}\right) \qquad (10.33)$$

FIGURE 10.9   Volkswagen 2.0 TDI diesel engine map.

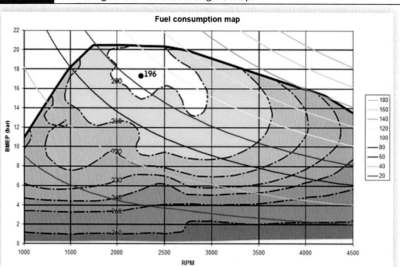

A bar is defined as 100kPa. In the engine map (Figure 10.10) the torque of the engine in ft-lb is equal to the bar reading on the y-axis multiplied by 11.74. A 2.0-l engine has a displacement volume of 0.002 m³.

$$\frac{\left(0.002\,\text{m}^3\right)\left(100{,}000\,\text{Pa/bar}\right)\left(0.73756\,\text{ft-lb/N}-\text{m}\right)}{4\pi} = \frac{11.74\,\text{ft-lb}}{\text{bar}} \qquad (10.34)$$

At points where the y-axis has a reading of 10 bar, the engine is producing 117.4 ft-lb of torque. The conversion factor will be different for different sizes of engines. For example, a 4.0 L engine will have a conversion factor twice that shown in Eq. 10.34. This illustrates how the BMEP approach can be used to compare engines of different sizes on the same graph. BMEP is used for gasoline engines too. It is not just for diesel engines.

Students sometimes ask why we do not use 2-cycle engines in cars and trucks. The advantage of 2-cycle engines is that they produce more horsepower for their weight than 4-cycle engines. Weed eaters, chain saws, and blowers used by lawn crews use 2-cycle engines because it makes them lighter in weight and weight is very important in those applications. Two-cycle engines inherently have a lower thermal efficiency than 4-cycle engines. A car with a 2-cycle engine would have the poor fuel economy. From an emissions standpoint, 2-cycle engines are filthy compared to 4-cycle engines. If all cars used 2-cycle engines, the air we breathe would be unbreathable. The 4-cycle engine is much cleaner, more efficient, less expensive to operate, and better in every way except that it is heavier than the 2-cycle engine.

Every engine will have a different mapping. The engineers who work out the details of the transmission design will fit the shift points of the transmission to optimize the efficiency of the engine, with a compromise to make the car drivable. For drivability we want the transmission to shift and allow the engine to make horsepower when we mash down on the gas pedal. It is hard to balance thermal efficiency and drivability. Optimizing too much for thermal efficiency can make the car seem sluggish to drive.

10.2 **Homework**

1. If 77.85 g of the gasoline fuel provides 1 kW-h of heat energy, what is the maximum thermal efficiency illustrated in the engine map (Figure 10.6)? [Answer: 31.8%]

2. Reading from the engine map (**Figure 10.10**):

    a. Assume that the engine provides 48 hp at 1200 RPM. What is the thermal efficiency? [Answer: 29.9%-30.3%]

    b. Assume that the engine provides 48 hp at 2200 RPM. What is the thermal efficiency? [Answer: 25.5%-26.4%]

    Please remember to calculate the torque so you can locate the (RPM, Torque) point on the graph. Use a ruler to help you locate the points on the graph. Each printer may scale the graph a little differently, and I won't know exactly how the graph will be scaled in the book. For my printer 388.5 ft-lb corresponded to 84.5 mm on the y-axis. Then knowing the ft-lb torque (T), the y-distance on the graph is calculated as y = (T*84.5/388.5). For the x-axis, 1000 RPM corresponded to zero and 4500 RPM corresponded to 145 mm. Then knowing the RPM (RPM), the x-distance as x = (RPM - 1000)*145/3500. Following this process allows for a little better accuracy using the graph.

FIGURE 10.10   Engine map homework #1.

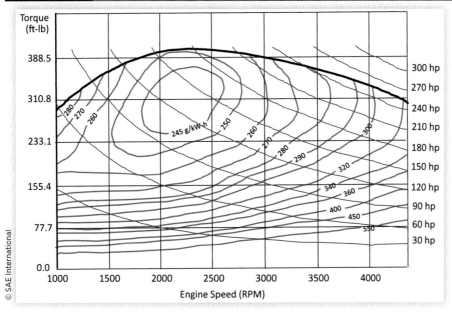

3. Assume that the engine powers an older large SUV that has a weight of 5400 lb, a rolling resistance coefficient of 0.012, a frontal area of 3.2 square meters, and a drag coefficient of 0.48. Assume that the transmission and final drive for the SUV are 88% efficient in getting the engine power to the wheels.

Assume that the vehicle is traveling on level ground at 45 mph. What is the ideal RPM for the engine to maximize fuel economy? Assume that the minimum RPM for the engine is 1000 RPM. Use the engine map to get the thermal efficiency of the engine and calculate the fuel economy of the SUV in mpg. [Answer: 25.9-26.9 mpg]

4. Operating the car at 1000 RPM will make the SUV feel sluggish. Suppose that the transmission causes the SUV in problem 3 to operate at 1500 RPM. Calculate the fuel economy for steady cruising at 45 mph and 1500 RPM. [Answer: 21.8-22.5 mpg]

5. Assume that the SUV in problems 3 and 4 is geared to operate at 2200 RPM at highway speed of 70 mph. Calculate the fuel economy of the SUV in mpg on the highway. [Answer: 15.3-15.4 mpg]

6. For this problem please neglect the rolling and aerodynamic power and calculate only the power required for acceleration. Assume that the weight of the vehicle is 4500 lbs.

   a. Assume that the vehicle is to accelerate from 0 to 30 mph at an acceleration rate of 12 ft/s². What is the maximum horsepower (at 30 mph) required for the vehicle?

   b. Assume that the vehicle is to accelerate from 60 mph to 80 mph at an acceleration rate of 6 ft/s2 to pass another vehicle on the highway. What is the maximum horsepower required (at 80 mph) for this acceleration? [Answer: 178.9 hp]

7. In this problem, you are to develop an acceleration spreadsheet like the one developed in Table 10.1 for the acceleration of the older large SUV in problems 3, 4, and 5. Assume that the acceleration is limited to 13 ft/s2 at lower speeds, and is limited by the horsepower of the engine at higher speeds. Assume that the engine will never provide more than 75% of the rated horsepower.

   a. Develop a graph of the speed of the vehicle vs time for a 30-s acceleration. Assume that the rated horsepower for the engine is 175 hp.

   b. What is the 0-60 mph time for the SUV? [Answer: 13.3 s]

   c. What is the time required for the SUV to accelerate from 60 mph to 80 mph?

   d. Develop a graph of the speed of the vehicle vs time for a 30-s acceleration. Assume that the rated horsepower for the engine is 275 hp.

   e. What is the 0-60 mph time for the SUV with the 275 hp engine?

   f. What is the time required for the SUV to accelerate from 60 mph to 80 mph with the 275 hp engine? [Answer: approximately 7.3 s]

11

Simulation of Fuel Economy for City, Suburban, and Highway Driving

11.1 Gasoline Engine Efficiency

Gasoline engines are typically 25%–30% efficient in converting the chemical energy in gasoline into useful mechanical energy when driving on the highway. The efficiency is zero when idling. At lower speeds, the efficiency is less than that on the highway. The engine is most efficient when operating at a specific revolution per minute (RPM) and is usually geared so that it is most efficient in high gear at 70 mph, with consideration for acceleration performance.

The ideal RPM for the engine depends on the amount of power required to push the vehicle along. When the vehicle requires a lot of power (like acceleration), it is more efficient to operate at a higher RPM than when it requires less power (cruising at low speed). Variable valve timing on modern engines has broadened the efficiency curve significantly and this allows the engines to operate more efficiently over a range of RPMs. Transmissions have lots of gears or are continuously variable. The electronics controlling the transmission will try to find the optimum gear ratio to allow the engine to operate at the most efficient RPM for the required power output. This sometimes creates a lag when you want to accelerate as the transmission must find a new gear to deliver the power, and drivers find the lag to be frustrating. But it provides a higher miles per gallon (MPG) fuel economy.

The graph (**Figure 11.1**) is data from fueleconomy.gov and is something to work with, but not a perfect model. It is for gasoline engines and typical mid-sized sedan cars built around the year 2000. New cars get better fuel economy at all speeds.

FIGURE 11.1 Overall fuel economy for cars.

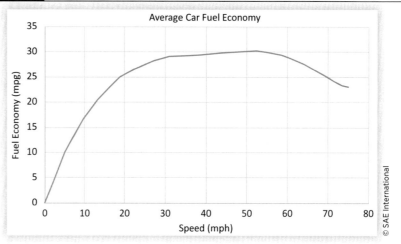

The power required will be the power to overcome rolling resistance and aerodynamic drag. In developing **Figure 11.2**, points were taken from Figure 11.1 for the fuel economy of the vehicle at different speeds. It was also assumed that the vehicle was a 4000-lb car with a coefficient of rolling resistance of Crr = 0.10, an aerodynamic drag area of 1.9 m², drag coefficient Cd = 0.33, and drive efficiency 90%. Figure 11.2 shows how the efficiency of the engine/drive varies with speed. This is an imperfect model, but it illustrates how the thermal efficiency of the gasoline engine varies with the overall average speed of the vehicle. For highway speeds, it works out best to assume a constant thermal efficiency of about 28%. For city driving, the thermal efficiency will be lower and depends on the power provided by the engine and the RPM as illustrated in the engine map. Figure 11.2 shows the overall thermal efficiency of the engine as a function of vehicle speed for an average car built around the year 2000.

FIGURE 11.2 Gasoline engine thermal efficiency at different speeds.

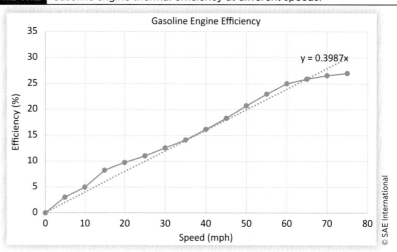

FIGURE 11.3   Thermal efficiency of Saturn 1.9L engine as a function of power.

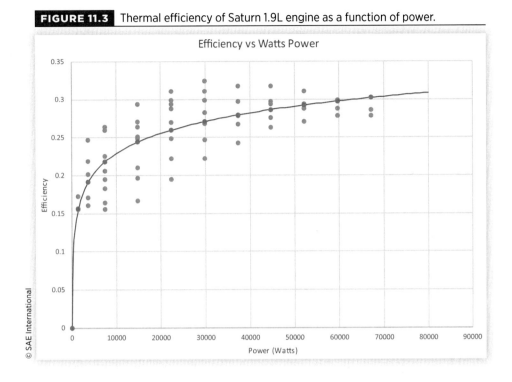

Looking at the engine map for the Saturn 1.9 L engine (Figure 10.1), it is clear the engine can produce power over a range of RPMs, and that the efficiency of the engine depends on the RPM as selected by the transmission ratio. In general, choosing a lower RPM (higher gear) allows the engine to operate at higher efficiency in producing the required power. Data were collected from the engine map (Figure 10.1) and presented in **Figure 11.3**.

On the x-axis of the graph is the power the engine is producing in watts. The y-axis shows the corresponding thermal efficiency. The blue dots on the graph are the data points collected from the engine map (Figure 10.1), and illustrate the engine operates over a range of thermal efficiencies for any required power output. The higher blue dots correspond to the power being produced at low RPMs (higher gears), and the lower blue dots on the graph correspond to the power being produced at higher RPMs (lower gears).

The red dots on the graph and the dotted line illustrate the model that will be used for modeling city and suburban driving. For city driving the transmission will choose a mid-range gear to produce the required power. If the transmission always chooses the highest possible gear for the engine to produce the required power the car will feel sluggish when the driver wants to accelerate. The driver will not be happy with the car. Choosing a mid-range gear as illustrated in the model will make the car more lively and fun to drive around town. The trend line through the data follows the curve efficiency = 0.0384 ln (power + 25.9)– 0.1249 for this engine.

To use the model for other gasoline-powered cars and trucks, it is necessary to normalize to the size of the engine. The model was developed for a 124 hp engine. To normalize the power term in the equation is multiplied by the quantity $(124/P_{max})$, where P_{max} is the rated hp of the engine. For city driving, we will use Eq. 11.1 to estimate the thermal efficiency of a gasoline engine:

$$\text{Thermal Efficiency} = (0.0384)\ln\left(\frac{124\,P}{P_{max}} + 25.9\right) - 0.1249 \qquad (11.1)$$

I apologize for what looks like a mix of units for P and P_{max}. The 124 in the equation has units of hp and to normalize it the P_{max} term needs to be the rating of the engine in hp. In the development that follows it is convenient to have the power produced by the engine in W, so the P term is in W.

a. P_{max} = Maximum rated horsepower of the engine in hp.

b. P = Power the engine is producing in W.

Internal combustion engines continue to use fuel when the car stops at a stoplight because they must be idle. Larger engines will use more fuel at idle than smaller engines. Including the idle power is a small but necessary correction for the models to be developed for city driving. For this analysis, we will assume the power at idle is $(0.002)P_{max}$, and the idle power will need to be converted to W.

$$\text{Idle Power} = (0.002)P_{max} \qquad (11.2)$$

The Saturn 1.9 L engine that the model is based on is rated at a maximum of 124 hp. Multiplying the power term in the log function by the ratio of $(124/P_{max})$ scales the model to different sizes of engines based on their power output capabilities. The assumption is that all gasoline engines will have an engine map similar to the Saturn 1.9L. Though not perfect, this is not a terrible assumption. Section 11.2 gives a more general approach if the reader has a specific car and engine in mind. Some gasoline engines will be more efficient than the Saturn 1.9 L, and some will be less efficient. I have tested the model on many production vehicles, and it has always yielded a reasonable estimate of the fuel economy as compared to the Environmental Protection Agency (EPA) estimate for city driving.

The model underestimates the fuel economy of vehicles for highway driving. When driving on the highway the transmission will shift to a higher gear and allow the car to be a little more sluggish to get better fuel economy. The thermal efficiency predicted by the model will be lower than what is achieved in highway driving. For highway driving, it is best to assume a constant thermal efficiency of about 28% for gasoline-powered cars.

Suburban driving is a mix of city and highway driving. Overall, the model used in this class is reasonably accurate in modeling suburban driving. In most cases, the car will get a slightly better fuel economy that is predicted by the model, but it is reasonably close.

11.2 General Approach to Model Engine Thermal Efficiency

To be able to model fuel economy for a car in city driving, it is necessary to have a model for the thermal efficiency of the engine as a function of the power it produces. We will use Eq. 11.1 for all of the examples in the book, but if the reader has a specific car and engine in mind, and if the reader has a good engine map, it is possible to develop a more precise equation to replace Eq. 11.1.

Start with an engine map similar to Figure 10.1. Follow the hp curves and take a data point each place it crosses a fuel consumption line. For example, looking at Figure 10.1, following the 30 hp curve on the graph starting on the right side of the graph and proceeding leftward, it crosses 475 g/kW-h, then 450, 425, 400, 375, 350, 325, 300, 275, 250. These values can be converted to thermal efficiency by taking 77.85 and dividing by the numbers above. The engine can produce 30 hp at all these different thermal efficiencies. Similar data can be taken by following the other hp curves. The data points can be plotted to yield a plot similar to the blue dots in Figure 11.3. A trend line is fit through the data points so that it falls in the middle of the data points. The trend line should have the form:

$$\text{Thermal Efficiency} = A \ln(P + B) + C \tag{11.3}$$

Start with A = 0.0384, B = 25.9, and C = −0.1249 since those numbers work for the Saturn engine. In the spreadsheet, take the absolute value of the difference between the trend line and each data point as the "error" and total the error. Adjust the parameters A, B, and C to minimize the total error, with the requirement that the trend line should go through (0,0). Since there are three parameters to adjust it will be impossible to know you have achieved the ideal optimal solution for their values, but it is possible to achieve a good solution. The trend line can then be used to estimate the thermal efficiency of the engine in city driving.

For highway driving, the transmission will shift to allow the engine to operate near the higher thermal efficiency points in the graph. The trend line will be a low estimate of the thermal efficiency for highway driving because it fits through the middle of the data. For highway driving, it is probably close enough to use a constant thermal efficiency for the engine as was done in unit 1 of the book. If the reader wanted to develop a formula for highway driving it would make sense to fit the trend line closer to the high values of thermal efficiency in Figure 11.3.

11.3 **Efficiency of Electric Motors**

The engine map in **Figure 11.4** is for a typical electric motor that would be used to power an electric vehicle.

The efficiency of an electric motor is good over a very large operating range. The efficiency is zero at zero RPMs. Above 10% of the rated RPMs and 10% of the rated torque the efficiency becomes very good. The transmission and drive systems will absorb some energy making the overall motor/drive system 85%–90% for an electric car. Solar cars and electrothon cars attach the wheel directly to the motor, avoiding the losses in the transmission and drive systems.

For modeling electric cars in city driving, we will assume an overall efficiency for the motor/drive system. The efficiency doesn't vary a lot, and this is an acceptable approximation.

FIGURE 11.4 Electric engine map.

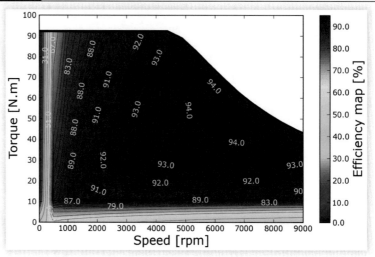

For hybrid electric cars, one of the main advantages is that we can turn the gasoline engine off when we stop and save the cost of idling. The electric motor is used to get the car started rolling and then the gasoline engine will start and provide the power for most of the driving. The small battery system is charged by the gasoline engine. All the energy for driving comes from the gasoline engine. The electric motor is used for low speed where it is much more efficient than the gasoline engine. The gasoline engine operates in a speed/power range where it is more efficient in powering the car and charging the battery. The result is better fuel economy for city driving.

11.4 **Simulation of Fuel Economy**

With the models for the efficiency of gasoline and electric motors, we can now begin modeling fuel economy for cars. The SAE J1082 specification was developed for road testing of vehicles in urban, suburban, and highway driving. We will develop mathematical models of vehicles driving the different cycles and use the results to compare fuel economy for different vehicles. The driving cycles are listed in the tables below.

SAE J1082 was developed for the actual driving of cars on a test track and it required a lot of practice for drivers to be able to perform the tasks listed in the tables. Results from the tests were dependent on the ability of the driver, and the specification is no longer used. But for our purposes, we will use the computer to simulate the driving in the specification and the computer will not have the issues of driver performance. There was a lot of thought put into developing the SAE J1082 specification and the driving patterns that would simulate urban, suburban, and highway driving. The specification is useful in simulating the fuel economy performance of cars in different types of driving (**Tables 11.1** through **11.4**).

TABLE 11.1 SAE J1082 urban cycle.

SAE J1082 Urban Driving Cycle	
Distance (Mile)	**Operation**
0.0	Start the fuel meter and timing device, idle 15 s, accelerate to 15 mph at 7 ft/s^2. Proceed at 15 mph to the 0.2-mi marker.
0.2	Stop at 4 ft/s^2, accelerate to 15 mph at 7 ft/s^2. Proceed at 15 mph to the 0.3-mi marker.
0.3	Decelerate to 5 mph at 4 ft/s^2, accelerate to 15 mph at 7 ft/s^2. Proceed at 15 mph to the 0.5-mi marker.
0.5	Stop at 4 ft/s^2, record fuel temperature and idle 15 s, accelerate to 20 mph at 7 ft/s^2. Proceed at 20 mph to the 0.7-mi marker.
0.7	Stop at 4 ft/s^2, accelerate to 20 mph at 7 ft/s^2. Proceed at 20 mph to the 0.8-m marker.
0.8	Decelerate to 10 mph at 4 ft/s^2, accelerate to 20 mph at 5 ft/s^2. Proceed at 20 mph to the 1.0-mi marker.
1.0	Stop at 4 ft/s^2, record fuel temperature and idle 15 s, accelerate to 15 mph at 7 ft/s^2, then to 25 mph at 5 ft/s^2. Proceed at 25 mph to the 1.2-mi marker.
1.2	Stop at 4 ft/s^2, accelerate to 15 mph at 7 ft/s^2, then to 25 mph at 5 ft/s^2. Proceed at 25 mph to the 1.3-mi marker.
1.3	Decelerate to 15 mph at 4 ft/s^2, accelerate to 25 mph at 5 ft/s^2. Proceed at 25 mph to the 1.5-mi marker.
1.5	Stop at 4 ft/s^2, record fuel temperature and idle 15 s, accelerate to 15 mph at 7 ft/s^2, then to 30 mph at 5 ft/s^2. Proceed at 30 mph to the 1.7-mi marker.
1.7	Stop at 4 ft/s^2, accelerate to 15 mph at 7 ft/s^2, and then to 30 mph at 5 ft/s^2. Proceed at 30 mph to the 1.8-mi marker.
1.8	Decelerate to 20 mph at 4 ft/s^2, accelerate to 30 mph at 5 ft/s^2. Proceed at 30 mph until it is time to begin braking for the 2.0-mi marker.
2.0	Begin braking at 4 ft/s^2 so the car will stop at the 2.0-mi marker. Stop fuel meter and timing device at stop, record fuel consumed, elapsed time, and fuel temperature.

TABLE 11.2 SAE J1082 suburban cycle.

SAE J1082 Suburban Driving Cycle	
Distance (Mile)	**Operation**
0.0	Approach the starting line at 40 mph. At the line start fuel measuring and timing devices, accelerate to 60 mph at 3 ft/s^2. Proceed at 60 mph to the 0.7-mi marker.
0.7	Decelerate to 30 mph at 4 ft/s^2. Accelerate to 50 mph at 3 ft/s^2. Proceed at 50 mph to the 2.0-mi marker.
2.0	Stop at 4 ft/s^2, record the fuel temperature and idle 7 s. Accelerate to 15 mph at 7 ft/s^2. Continue accelerating to 25 mph at 5 ft/s^2. Continue accelerating to 40 mph at 3 ft/s^2. Proceed at 40 mph to the 2.6-mi marker.
2.6	Accelerate to 50 mph at 3 ft/s^2. Proceed at 50 mph to the 3.3-mi marker.
3.3	Stop at 4 ft/s^2, record the fuel temperature and idle 7 s. Accelerate to 15 mph at 7 ft/s^2. Continue accelerating to 25 mph at 5 ft/s^2. Continue accelerating to 40 mph at 3 ft/s^2. Proceed at 40 mph to the 5.2-mi marker.
5.2	Stop the fuel measuring and timing devices while driving at 40 mph at the 5.2-mi marker. Record fuel consumed, elapsed time, and fuel temperature.

TABLE 11.3 SAE J1082 Interstate cycle (55 mph).

SAE J1082 Interstate Cycle (55 mph)	
Distance (Mile)	**Operation**
0.0	Approach the starting line at 55 mph. Record fuel temperature at the line and start fuel measuring and timing devices. Proceed at 55 mph to the 0.2-mi marker.
0.2	Accelerate to 60 mph at 1 ft/s². Immediately decelerate to 50 mph at 1 ft/s². Immediately accelerate to 55 mph at 1 ft/s². Proceed at 55 mph to the 1.2-mi marker.
1.2	Accelerate to 60 mph at 1 ft/s². Immediately decelerate to 50 mph at 1 ft/s². Immediately accelerate to 55 mph at 1 ft/s². Proceed at 55 mph to the 2.2-mimarker.
2.2	Accelerate to 60 mph at 1 ft/s². Immediately decelerate to 50 mph at 1 ft/s². Immediately accelerate to 55 mph at 1 ft/s². Proceed at 55 mph to the 3.2-mi marker.
3.2	Accelerate to 60 mph at 1 ft/s². Immediately decelerate to 50 mph at 1 ft/s². Immediately accelerate to 55 mph at 1 ft/s². Proceed at 55 mph to the 4.7-mi marker.
4.7	Stop the fuel measuring and timing devices while driving at 55 mph at the 4.7-mimarker. Record fuel consumed, elapsed time, and fuel temperature.

TABLE 11.4 SAE J1082 interstate cycle (70 mph).

SAE J1082 Interstate Cycle (70 mph)	
Distance (Mile)	**Operation**
0.0	Approach the starting line at 70 mph. Record fuel temperature at the line and start fuel measuring and timing devices. Proceed at 70 mph to the 0.2-mi marker.
0.2	Accelerate to 75 mph at 1 ft/s². Immediately decelerate to 65 mph at 1 ft/s². Immediately accelerate to 70 mph at 1 ft/s². Proceed at 70 mph to the 1.2-mi marker.
1.2	Accelerate to 75 mph at 1 ft/s². Immediately decelerate to 65 mph at 1 ft/s². Immediately accelerate to 70 mph at 1 ft/s². Proceed at 70 mph to the 2.2-mi marker.
2.2	Accelerate to 75 mph at 1 ft/s². Immediately decelerate to 65 mph at 1 ft/s². Immediately accelerate to 70 mph at 1 ft/s². Proceed at 70 mph to the 3.2=mi marker.
3.2	Accelerate to 75 mph at 1 ft/s². Immediately decelerate to 65 mph at 1 ft/s². Immediately accelerate to 70 mph at 1 ft/s². Proceed at 70 mph to the 4.7-mi marker.
4.7	Stop the fuel measuring and timing devices while driving at 70 mph at the 4.7-mi marker. Record fuel consumed, elapsed time, and fuel temperature.

For these models, we will assume the road is flat so that gravitational energy is not a factor. The car will accelerate and decelerate, which means that the kinetic energy of the car will change. The change in kinetic energy between steps i and i+1 is:

$$\text{Change in Kinetic Energy} = \frac{1}{2}m\left(V_{i+1}^2 - V_i^2\right) \tag{11.4}$$

If the change in kinetic energy happens over a time step T, the average kinetic power over that time step is:

$$\text{Kinetic Power} = \frac{\frac{1}{2}m\left(V_{i+1}^2 - V_i^2\right)}{T} \tag{11.5}$$

If the weight of the vehicle (W) is in pounds and the speed (V) is in mph there will be unit conversions to get the energy and power into sensible units. It is assumed the time step T used in the spreadsheet is in seconds. The following equations are useful in developing the models. In all equations, it is assumed the vehicle weight is in lb, the speed in mph, and the aerodynamic drag area in m². Several of the equations are repeated from Unit 1 for convenience.

$$\text{Kinetic Energy} = (0.0454)\,WV^2\ J \tag{11.6}$$

$$\text{Kinetic Energy} = (0.0334)\,WV^2\ \text{ft-lb} \tag{11.7}$$

$$\text{Kinetic Power} = \frac{(0.0454)\,W\left(V_{i+1}^2 - V_i^2\right)}{T}\ W \tag{11.8}$$

$$\text{Kinetic Power} = \frac{(0.0334)\,W\left(V_{i+1}^2 - V_i^2\right)}{T}\ \text{ft-lb/s} \tag{11.9}$$

$$\text{Rolling Power} = (1.989)\,W\,V\left(1 + \frac{V}{200}\right)\ W \tag{11.10}$$

$$\text{Rolling Power} = (1.467)\,W\,V\left(1 + \frac{V}{200}\right)\ \text{ft-lb/s} \tag{11.11}$$

$$\text{Aero Power} = (0.05357)\,V^3\ A\ Cd\ W \tag{11.12}$$

$$\text{Aero Power} = (0.03931)\,V^3\ A\ Cd\ \text{ft-lb/s} \tag{11.13}$$

If the road is flat so that there is no change in gravitational energy, the total power required from the batteries for electric vehicles, or the fuel for internal combustion engine vehicles is:

$$\text{Total Power} = \frac{\text{Kinetic Power} + \text{Rolling Power} + \text{Aero Power}}{\text{motor/drive efficiency}} \tag{11.14}$$

Electric Vehicle Model: The easiest model to develop for city driving is the electric vehicle model, so it will be developed first. For the electric vehicle, we will assume that the motor/drive efficiency is constant.

Example 11.1: Model an electric car driving the urban cycle. Assume that there is no regenerative braking for this case. Parameters for the car are: W = 3800 lb, Crr = 0.011, A = 1.9 m², Cd = 0.32, motor/drive efficiency = 85%, battery capacity = 30,000 W-h. This example was developed by using parameters for the 2017 Nissan Leaf electric car.

1. Determine the average W-h per mi energy consumption for urban driving.

2. What is the range of the vehicle for urban driving?

Solution: We start with modeling an electric car because it is easier than modeling a gasoline or diesel-powered car. The W-h of electric energy used per mi traveled can be related to the cost per mi of driving the car by multiplying by the cost of the electricity used to charge the batteries. The cost of the electricity per mi for an electric car is significantly less than the cost of gasoline per mi for a gasoline-powered car. The more critical factor for an electric car is the driving range.

TABLE 11.5 Urban driving for an electric car.

Time (s)	Acc (ft/s²)	Speed (ft/s)	Speed (mph)	Dist (ft)	Dist (miles)	Watts (roll)	Watts (aero)	Watts (kin)	Watts (total)	Watts (clean)	Watt-hrs
16	7	0	0	0	0	0	0	0	0	0	0
17	7	7	4.772	3.5	0.000663	406.3	3.54	3930	5105	5105	0.709
18	7	14	9.545	14	0.002652	831.5	28.3	11,780	14,882	14,882	3.485
19	1	21	14.32	31.5	0.005966	1276	95.6	19,649	24,730	24,730	8.987
20	0	22	15	53	0.010038	1341	109.9	3449	5764	5764	13.22

© SAE International

Looking at the Urban Cycle, the car idles for the first 15 s, simulating being stopped at a stoplight. An electric car will be off and consuming no energy when at the stoplight, so the energy consumed in the first 15 s is zero. **Table 11.5** is used to illustrate how the model was developed.

The time step chosen for this model was 1 s. The values in all the other columns are zero for the first 15 s and the table above starts at time 16 s. At the beginning of the 16th s, the car accelerates at 7 ft/s². The speeds and power are all zero at the beginning of the time step. The way the model is developed, each row represents what is happening at the beginning of the time step.

In column two, the car accelerates at 7 ft/s² until it reaches 15 mph, which is 22 ft/s. The speed in ft/s is shown in the third column and converted to mph in the fourth column. Notice that at time step 19 s the acceleration is reduced to 1 ft/s² because that is what is needed to make the speed 15 mph at the beginning of time step 20. The first 4 columns of the spreadsheet are developed using Table 11.1 so that the car has the correct speed and acceleration according to the specification SAE J1082.

Column 5 is the distance traveled by car. Distance is the average velocity during the time step multiplied by the time step (T =1 s in this case). Using D for the distance in column 5 and V for the speed in column 3 the distances (ft) are calculated as:

$$D_{i+1} = D_i + \frac{1}{2}(V_{i+1} + V_i)(T) \tag{11.15}$$

The distances are converted to mi in column 6. The first 6 columns are the same for any vehicle traveling on the urban cycle. None of the terms depend on the weight of the vehicle, the rolling or aerodynamic losses, or other car parameters. The columns on the right side of the table are used to estimate the amount of electric energy used during each time step and then total it for the cycle.

Columns 7, 8, 9, and 10 calculate the rolling, aero, and kinetic power and then total the power. To be sure you understand how to do the calculations you should go through the calculations and make sure the numbers match what is in the table. The calculations are done using Eqs. 11.8, 11.10, and 11.12 to calculate rolling, aero, and kinetic power. The total power is calculated in column 10 using Eq. 11.16.

$$\text{Total Power} = \frac{\text{Rolling} + \text{Aero} + \text{Kinetic}}{\text{motor} / \text{drive efficiency}} \tag{11.16}$$

A logic statement is required for column 11 because there are times in the spreadsheet where the car is braking. Braking will yield a negative number for the kinetic power and for the total power. For the first example we are assuming there is no regenerative braking, so all the braking energy will be lost as heat in the brakes. The total power is positive for rows 16–20, but there will be places in the spreadsheet where the total power is negative. Column 11 is the "cleaned" power which will zero the power requirements when the total power is negative. In Excel the logic statement is:

$$\text{If}\left(\text{Total Power} > 0, \text{Total Power}, 0\right) \tag{11.17}$$

In the next example, we will include regenerative braking and assess how much difference regenerative braking makes in extending the range of the vehicle. This logic statement (Eq. 11.17) will be modified to include regenerative braking.

The last column in the spreadsheet totals the number of watt-h energy used from the batteries. The energy used is the average power during the time step multiplied by the time step. The 3600 factor in the equation converts the units from joules to watt-hrs. Let P be the cleaned power in column 11, and E be the energy used in column 12. The total energy used in column 12 is calculated from Eq. 11.18:

$$E_{i+1} = E_i + \frac{\frac{1}{2}\left(P_{i+1} + P_i\right)\left(T\right)}{3600} \tag{11.18}$$

The reader should go through enough calculations to feel confident with the equations. The model can then be developed as shown in **Table 11.6**. The actual spreadsheet goes to a time of 480 s. Table 11.6 shows the first 26 s of the urban simulation. Readers can use the numbers in Table 11.6 to check and make sure the spreadsheet they are developing is working correctly.

A portion of the spreadsheet where the total power goes negative is shown in **Table 11.7** to illustrate how the logic statement in column 11 works when the car is braking.

In column 2 notice that the car is braking at a deceleration of −4 ft/s² during the first part of the table. The kinetic power in column 10 goes negative when the car is braking. The "cleaned" power column sets the negative power to zero because without regenerative braking the car cannot recover the braking energy. The total energy used from the batteries in the far right column does not increase during braking. In the next example, we will add a factor for regenerative braking and allow the car to recover some of the braking energy. When we include regenerative braking the total energy used from the batteries in the far right column will decrease slightly as energy is put back into the batteries.

The number at the bottom of column 6 (at 480 s) is the total distance traveled by car in mi. The number at the bottom of column 12 is the total number of W-h of energy used from the batteries. Dividing the total W-h used by the total distance traveled gives the 254.2 W-h/mi reading shown in the top right corner of the spreadsheet. The range of the vehicle is calculated as the battery capacity divided by the 254.2 W-h/mi and is 118 miles in this case.

TABLE 11.6 Urban driving model for an electric car.

Car Weight	3800	lb		Battery	30000	W-h				W-h/mi	254.2	
Crr	0.011			eff	0.85					Range	118.0	
Area	1.9	m²								Regen eff	0	
Cd	0.32											
s	ft/s^2	ft/s	mph	feet	Mile		Watt	Watt	Watt	Watt	Watt	W-h
Time	Acceleration	Speed	Speed	Distance	Distance		(roll)	(aero)	(kinetic)	(total)	(cleaned)	Total
0	0	0	0	0	0		0	0	0	0	0	0
1	0	0	0	0	0		0	0	0	0	0	0
2	0	0	0	0	0		0	0	0	0	0	0
3	0	0	0	0	0		0	0	0	0	0	0
4	0	0	0	0	0		0	0	0	0	0	0
5	0	0	0	0	0		0	0	0	0	0	0
6	0	0	0	0	0		0	0	0	0	0	0
7	0	0	0	0	0		0	0	0	0	0	0
8	0	0	0	0	0		0	0	0	0	0	0
9	0	0	0	0	0		0	0	0	0	0	0
10	0	0	0	0	0		0	0	0	0	0	0
11	0	0	0	0	0		0	0	0	0	0	0
12	0	0	0	0	0		0	0	0	0	0	0
13	0	0	0	0	0		0	0	0	0	0	0
14	0	0	0	0	0		0	0	0	0	0	0
15	0	0	0	0	0		0	0	0	0	0	0
16	7	0	0	0	0		0	0	0	0	0	0
17	7	7	4.77	3.5	0.00066		406.3	3.54	3930	5105	5105	0.709
18	7	14	9.55	14	0.00265		831.5	28.33	11789	14882	14882	3.485
19	1	21	14.32	31.5	0.00597		1275.6	95.61	19649	24730	24730	8.987
20	0	22	15	53	0.01004		1340.6	109.93	3449	5764	5764	13.222
21	0	22	15	75	0.01420		1340.6	109.93	0	1707	1707	14.259
22	0	22	15	97	0.01837		1340.6	109.93	0	1707	1707	14.733
23	0	22	15	119	0.02254		1340.6	109.93	0	1707	1707	15.207
24	0	22	15	141	0.02670		1340.6	109.93	0	1707	1707	15.682
25	0	22	15	163	0.03087		1340.6	109.93	0	1707	1707	16.156
26	0	22	15	185	0.03504		1340.6	109.93	0	1707	1707	16.630

The simulation shows that the electric car would have a range of 118 miles in city driving. If the electricity to charge the batteries cost 10 cents per kW-h, the cost to charge the batteries would be $3.00 and the cost per mile would be about $0.025/mi. If we compare to a gasoline-powered car that gets 30 mpg in the city and burns fuel that costs $2.50 per gallon the fuel cost is about $0.083/mi. The energy cost of driving an electric car is much less than for a similar size gasoline power car.

TABLE 11.7 Portion of an urban driving model for electric cars illustrating what happens during braking.

s Time	ft/s^2 Acceleration	ft/s Speed	mph Speed	feet Distance	Mile Distance	Watt (roll)	Watt (aero)	Watt (kinetic)	Watt (total)	Watt (cleaned)	W-h Total
65	0	22	15.00	1043	0.19754	1340.6	109.93	0	1707	1707	35.117
66	−4	22	15.00	1065	0.20170	1340.6	109.93	0	1707	1707	35.591
67	−4	18	12.27	1085	0.20549	1083.0	60.21	−12832	−13752	0	35.828
68	−4	14	9.55	1101	0.20852	831.5	28.33	−10266	−11066	0	35.828
69	−4	10	6.82	1113	0.21080	586.2	10.32	−7699	−8356	0	35.828
70	−4	6	4.09	1121	0.21231	347.1	2.23	−5133	−5628	0	35.828
71	−2	2	1.36	1125	0.21307	114.1	0.08	−2566	−2885	0	35.828
72	7	0	0.00	1126	0.21326	0.0	0.00	−321	−377	0	35.828
73	7	7	4.77	1129.5	0.21392	406.3	3.54	3930	5105	5105	36.537
74	7	14	9.55	1140	0.21591	831.5	28.33	11789	14882	14882	39.313
75	1	21	14.32	1157.5	0.21922	1275.6	95.61	19649	24730	24730	44.815
76	0	22	15.00	1179	0.22330	1340.6	109.93	3449	5764	5764	49.050

The range of the electric car is the main issue. Everyone would drive electric cars if the cost was comparable to gasoline-powered cars and if they had adequate range. Getting an adequate range for the car has always been a problem. One of the things that can be done to improve the range in city driving is to put regenerative braking on the car. Then when the car brakes some of the braking energy can be put back into the batteries and the range of the car will be extended. Safety considerations prevent us from being able to put all the braking energy back into the batteries. It is more important to be able to stop the car in a controlled manner than to recover the braking energy. Realistically it is possible to recover about 30% of the braking energy using a regenerative braking system. To accommodate this in the spreadsheet we allow a fraction of the braking energy to be put back in the batteries and recognize that 30% is as good as we will do in recovering the braking energy. The logic statement in column 11 is modified to be:

$$\text{If}\left(\text{Total Power} > 0, \text{Total Power}, \text{regen eff} * \text{Total Power}\right) \qquad (11.19)$$

Example 11.2: Modify the spreadsheet in the previous example to illustrate how regenerative braking effects the efficiency and range of the electric car, with the regenerative braking efficiency set to 30%.

Solution: The "cleaned" power column is modified using Eq. 11.19. The result is that the average energy consumption per mi is reduced to 215.3 W-h/mi and the range is extended to 139.4 mi. Being able to recover 30% of the braking energy increases the range of the car from 118 mi to 139.4 mi: a very significant increase. The first 26 s of the model are shown in **Table 11.8** to allow readers who are developing spreadsheets to compare numbers.

TABLE 11.8 Urban driving for an electric car with regenerative braking.

Car Weight	3800	lb		Battery	30000	W-h				W-h/mi	215.3	
Crr	0.011			eff	0.85					Range	139.4	
Area	1.9	m²								Regen eff	0.3	
Cd	0.32											
s	ft/s^2	ft/s	mph	feet	Mile		Watt	Watt	Watt	Watt	Watt	W-h
Time	Acceleration	Speed	Speed	Distance	Distance		(roll)	(aero)	(kinetic)	(total)	(cleaned)	Total
0	0	0	0	0	0		0	0	0	0	0	0
1	0	0	0	0	0		0	0	0	0	0	0
2	0	0	0	0	0		0	0	0	0	0	0
3	0	0	0	0	0		0	0	0	0	0	0
4	0	0	0	0	0		0	0	0	0	0	0
5	0	0	0	0	0		0	0	0	0	0	0
6	0	0	0	0	0		0	0	0	0	0	0
7	0	0	0	0	0		0	0	0	0	0	0
8	0	0	0	0	0		0	0	0	0	0	0
9	0	0	0	0	0		0	0	0	0	0	0
10	0	0	0	0	0		0	0	0	0	0	0
11	0	0	0	0	0		0	0	0	0	0	0
12	0	0	0	0	0		0	0	0	0	0	0
13	0	0	0	0	0		0	0	0	0	0	0
14	0	0	0	0	0		0	0	0	0	0	0
15	0	0	0	0	0		0	0	0	0	0	0
16	7	0	0	0	0		0	0	0	0	0	0
17	7	7	4.77	3.5	0.00066		406.3	3.54	3930	5105	5105	0.709
18	7	14	9.55	14	0.00265		831.5	28.33	11789	14882	14882	3.485
19	1	21	14.32	31.5	0.00597		1275.6	95.61	19649	24730	24730	8.987
20	0	22	15	53	0.01004		1340.6	109.93	3449	5764	5764	13.222
21	0	22	15	75	0.01420		1340.6	109.93	0	1707	1707	14.259
22	0	22	15	97	0.01837		1340.6	109.93	0	1707	1707	14.733
23	0	22	15	119	0.02254		1340.6	109.93	0	1707	1707	15.207
24	0	22	15	141	0.02670		1340.6	109.93	0	1707	1707	15.682
25	0	22	15	163	0.03087		1340.6	109.93	0	1707	1707	16.156
26	0	22	15	185	0.03504		1340.6	109.93	0	1707	1707	16.630

A portion of the spreadsheet illustrating what happens during regenerative braking is shown in **Table 11.9**.

Table 11.9 Portion of urban driving model for electric cars illustrating what happens during regenerative braking. Notice that during regenerative braking the total W-h drawn from the batteries in the right column decreases slightly because energy is being put into the batteries. During this regenerative braking cycle, the energy drawn from

TABLE 11.9 Portion of an urban driving model for electric cars illustrating what happens during regenerative braking.

s Time	ft/s^2 Acceleration	ft/s Speed	mph Speed	feet Distance	Mile Distance	Watt (roll)	Watt (aero)	Watt (kinetic)	Watt (total)	Watt (cleaned)	W-h Total
65	0	22	15.00	1043	0.19754	1340.6	109.93	0	1707	1707	35.117
66	−4	22	15.00	1065	0.20170	1340.6	109.93	0	1707	1707	35.591
67	−4	18	12.27	1085	0.20549	1083.0	60.21	−12832	−13752	−4125	35.255
68	−4	14	9.55	1101	0.20852	831.5	28.33	−10266	−11066	−3320	34.221
69	−4	10	6.82	1113	0.21080	586.2	10.32	−7699	−8356	−2507	33.412
70	−4	6	4.09	1121	0.21231	347.1	2.23	−5133	−5628	−1688	32.829
71	−2	2	1.36	1125	0.21307	114.1	0.08	−2566	−2885	−865	32.475
72	7	0	0.00	1126	0.21326	0.0	0.00	−321	−377	−113	32.339
73	7	7	4.77	1129.5	0.21392	406.3	3.54	3930	5105	5105	33.032
74	7	14	9.55	1140	0.21591	831.5	28.33	11789	14882	14882	35.808
75	1	21	14.32	1157.5	0.21922	1275.6	95.61	19649	24730	24730	41.310
76	0	22	15.00	1179	0.22330	1340.6	109.93	3449	5764	5764	45.545

the batteries goes from 35.591 W-h down to 32.339 W-h, indicating that 3.252 W-h of energy was recovered.

The model shows that adding regenerative braking to the car will extend the range from 118 miles to 139.4 miles, an improvement of 18%. Once the model is developed we can use it to help us understand how the different car parameters impact the range of the vehicle. We know that reducing the weight of the car will give the car a better range in the city. The model allows us to quantify how much improvement in the range can be achieved.

Example 11.3: Suppose we consider replacing some of the steel components in the car with aluminum and estimate it will reduce the weight of the car by 100 lb. How much does this impact the range of the vehicle?

Solution: Change the weight in the top left corner to 3700 and the spreadsheet will automatically recalculate everything. The range of the vehicle improves to 142.85 mi. Reducing the weight by 100 lb will improve the range by 3.5 mi. There will be a cost associated with replacing the steel components with aluminum and we will need to decide if the additional 3.5 miles in the range is worth the additional cost.

Example 11.4: Suppose we find better tires for the car that have a rolling resistance coefficient of 0.009 rather than the 0.011 of the tires in the model. Changing the Crr to 0.009 increases the range of the vehicle to 152.04 miles, an improvement of about 12.5 miles. From the analysis, it is clear that using low rolling resistance tires is important for electric vehicles in city traffic.

Example 11.5: Suppose we work on the aerodynamics of the car and reduce the aerodynamic drag coefficient to 0.30. Making that change in the model causes the range of the vehicle to improve to 140.00 miles, an improvement of 0.64 miles. This analysis shows that aerodynamics is not very important for city driving. When we model the car in the highway driving cycle the aerodynamic drag will be more important, but for city driving it is not very important.

Example 11.6: For another trade-off consider what happens by adding more batteries. Suppose adding 70 lb of batteries increases the battery capacity by 4000 W-h. This would be approximately correct for adding lithium-ion batteries to the car. Adding 70 lb to the weight of the car and 4000 W-h to the battery capacity improves the range to 155.29 mi, which is an improvement of about 16 miles in range. It also adds cost to the car and the batteries will take up some of the storage space in the car. The designer will need to decide if it is worth adding the additional batteries.

The models help guide the design of the car. We want to spend development money and effort on things that will help most in the performance of the car, and for an electric vehicle range is one of the most important objectives of the design. There are many possibilities for improving the range of the electric car, and nearly all of them involve added cost. Adding regenerative braking improved the range of the car 18%. It is almost certain that the added cost of regenerative braking is well spent. Reducing the weight of the vehicle using aluminum components is of less benefit. It may or may not be worth it. The model allows the designer to quantify how much improvement in the range is associated with different design alternatives and is an important design tool.

Gasoline Vehicle Model: The next step is to model a gasoline car in city driving. The process is very similar to the electric vehicle. Eq. 11.1 will be used to estimate the thermal efficiency of the engine and Eq. 11.2 will be used to estimate the power the engine consumes while idling.

Example 11.7: Assume a pickup has a weight of 5500 lb and the tires have a rolling resistance coefficient Crr = 0.010. The power rating for the engine is 375 hp. The frontal area is 2.77 m² and the drag coefficient is Cd = 0.50. Assume the power associated with idling is given by Eq. 11.2 (0.002*P_{max}). The drive efficiency is 90%. This example is developed using specifications for the 2018 Ford F-150 truck.

Solution: For this model, we can start with the model for the electric car. The rolling, aerodynamic, kinetic, and total power columns are calculated the same way as for the electric car. The truck is a larger and heavier vehicle, so the numbers are larger, but the formulas used are the same. When we do the "cleaned" column after the total power we need to recognize that the minimum power the engine provides is idle. The logic statement will need to say that if the total power is less than the idle power, then the truck is using idle power. Idle power for the truck is estimated as:

$$\text{Idle Power} = (0.002)(375)(745.7) = 559 \text{ W} \tag{11.20}$$

TABLE 11.10 Urban driving for a gasoline vehicle.

Time (s)	Acc (ft/s²)	Speed (ft/s)	Speed (mph)	Dist (ft)	Dist (miles)	Watts (roll)	Watts (aero)	Watts (kin)	Watts (total)	Watts (clean)
16	7	0	0	0	0	0	0	0	0	559
17	7	7	4.772	3.5	0.000663	534.6	8.06	5688	6923	6923
18	7	14	9.545	14	0.002652	1094	64.5	17,064	20,247	20,247
19	1	21	14.32	31.5	0.005966	1678	218	28,439	33,706	33,706
20	0	22	15	53	0.010038	1764	250	4991	7784	7784

The truck is stationary at the stoplight for the first 15 s, so the power requirement is idle power through the beginning of the 16th step. As with the electric car the first 15 steps are omitted in the portion of the spreadsheet shown in the **Table 11.10**. The reader should work through the calculations to ensure understanding.

There are too many columns in this spreadsheet to fit everything on one-page width. The print becomes too small. Added to the right of the table above will be 4 more columns used to calculate the amount of fuel used as the truck goes through the urban driving model. The four columns are below (**Table 11.11**):

Once we know the power the engine must produce, we can use the efficiency formula to calculate the thermal efficiency of the engine. Remember that P in Eq. 11.1 is the required engine power in watts and P_{max} is the rated engine power in horsepower, which in this case is 375 hp. The first column in Table 11.11 is calculated using Eq. 11.21, where P is from the far-right column in Table 11.10.

$$\text{Efficiency} = (0.0384)\ln\left[\frac{124\,P}{375} + 25.9\right] - 0.1249 \qquad (11.21)$$

The fuel power in the second column of Table 11.11 is equal to the engine power divided by the efficiency as shown in Eq. 11.22.

$$\text{Fuel Power} = \frac{P}{\text{Efficiency}}\,W \qquad (11.22)$$

Since there are 1055 J in one BTU, the fuel power can be converted to BTU/sec by dividing by 1055 as shown in column 3 of **Table 11.11**. Assuming 121,000 BTU/gallon, the fuel used during the time step is the BTU/s fuel power divided by 121,000. This result

TABLE 11.11 Efficiency and fuel consumption of gasoline vehicles in urban driving.

Efficiency	Watts (fuel)	BTU/s (fuel)	Gallons (fuel)
0.08058	6941	6.579	0.0000544
0.17259	40,111	38.02	0.000369
0.21352	94,826	89.88	0.001111
0.23303	144,643	137.10	0.002244
0.17705	43,967	41.67	0.002589

is multiplied by the time step T, which in this case is 1 s. We want to keep a running total in the gallons of fuel used in the fourth column of Table 11.11.

$$\text{Gallons}_{i+1} = \text{Gallons}_i + \frac{(\text{BTU}/\sec_{i+1})(1\,\text{s})}{121{,}000} \tag{11.23}$$

The reader should be able to calculate the numbers in the tables using the formulas given. The distance traveled in miles is at the bottom of the distance column and the gallons of gasoline used are at the bottom of the gallon's column. Dividing the miles traveled by the gallons used gives the average fuel economy of the truck in city traffic. The EPA estimate for the F-150 truck is 18 mpg in the city. The results of the model show the fuel economy to be 18.073 mpg. The print is small in **Table 11.12**, but it shows the spreadsheet for the first 30 s of the urban gasoline vehicle model.

Ford has gone through considerable effort to lighten the truck by using aluminum body components. If the weight of the vehicle is reduced by 500 lb, the fuel economy improves from 18.073 mpg to 19.228 mpg, which is a 6.4% improvement in fuel economy. Other studies can be performed to see how changing other car parameters impacts the overall fuel economy of the truck in city driving.

Many cars now have an instantaneous fuel economy meter on the dashboard that tells you your instantaneous fuel economy while driving. The instantaneous fuel economy is the distance traveled divided by the fuel used over a brief time period. We can plot this in the model. The plot is shown in **Figure 11.5** below.

Examining Figure 11.5 shows that the fuel economy is zero when idling. Not much fuel is consumed when idling, so idling does not have a huge impact on the overall fuel economy. The fuel economy is low during hard acceleration and significant amounts of fuel are consumed. Fuel economy levels out at 31 mpg when the truck is at a constant speed in this model because we are assuming it all takes place on level pavement. If small grades were added in to make the model more like a city driving the constant regions shown in the chart would not be constant. The high mpg readings on the chart happen when the truck is braking. The instantaneous fuel economy varies from zero to 150 mpg, but the average fuel economy is 18 mpg. The instantaneous fuel meter is interesting to watch, but I'm not sure the information is meaningful.

Hybrid Electric Vehicles: One of the main disadvantages of an internal combustion engine car is that it burns fuel while idling at stoplights or any time the cars stop and is not shut off. It is a waste of fuel and it lowers the fuel economy of the vehicle. There have been attempts with conventional gas-powered cars to simply have the car shut off when it stops and then automatically start when the driver presses on the gas pedal. This would result in significant fuel savings in city driving, but the controls to make this happen smoothly have been elusive. If we could overcome the control problems this would be an inexpensive way to improve the fuel economy in city driving. This can be modeled for the F-150 pickup in the previous example by setting the idle power to zero rather than the $(0.002)P_{max}$. Making this change in the spreadsheet improves the fuel economy from 18 mpg to 19.35 mpg, which is a significant improvement for such a simple modification.

Another problem with the internal combustion engine cars is that the thermal efficiency of the engine is low at low speeds or low power requirements, as when taking

TABLE 11.12 Urban model for a gasoline vehicle.

Truck Weight	5500	lb		Pmax	375	hp			Fuel Energy	121000	BTU/gal			
Crr	0.01			eff	0.9									
Area	2.77	m^2		idle	559	W			Econ	18.073	mpg			
Cd	0.5													
s	ft/s^2	ft/s	mph	feet	Miles	W	W	W	W	W	Thermal	W	BTU/s	gallon
Time	Acceleration	Speed	Speed	Dist	Dist	Rolling	Aero	Kinetic	Total	Clean	Efficiency	Fuel	Fuel	fuel
0	0	0	0	0	0	0	0	0	0	559	0.081	6941	6.58	0.00000
1	0	0	0	0	0	0	0	0	0	559	0.081	6941	6.58	0.00005
2	0	0	0	0	0	0	0	0	0	559	0.081	6941	6.58	0.00011
3	0	0	0	0	0	0	0	0	0	559	0.081	6941	6.58	0.00016
4	0	0	0	0	0	0	0	0	0	559	0.081	6941	6.58	0.00022
5	0	0	0	0	0	0	0	0	0	559	0.081	6941	6.58	0.00027
6	0	0	0	0	0	0	0	0	0	559	0.081	6941	6.58	0.00033
7	0	0	0	0	0	0	0	0	0	559	0.081	6941	6.58	0.00038
8	0	0	0	0	0	0	0	0	0	559	0.081	6941	6.58	0.00043
9	0	0	0	0	0	0	0	0	0	559	0.081	6941	6.58	0.00049
10	0	0	0	0	0	0	0	0	0	559	0.081	6941	6.58	0.00054
11	0	0	0	0	0	0	0	0	0	559	0.081	6941	6.58	0.00060
12	0	0	0	0	0	0	0	0	0	559	0.081	6941	6.58	0.00065
13	0	0	0	0	0	0	0	0	0	559	0.081	6941	6.58	0.00071
14	0	0	0	0	0	0	0	0	0	559	0.081	6941	6.58	0.00076
15	0	0	0	0	0	0	0	0	0	559	0.081	6941	6.58	0.00082
16	7	0	0	0	0	0	0	0	0	559	0.081	6941	6.58	0.00087
17	7	7	4.77	3.5	0.0007	535	8	5688	6923	6923	0.173	40111	38.02	0.00118
18	7	14	9.55	14	0.0027	1094	65	17064	20247	20247	0.214	94826	89.88	0.00193
19	1	21	14.32	31.5	0.0060	1678	218	28439	33706	33706	0.233	144643	137.10	0.00306
20	0	22	15	53	0.0100	1764	250	4991	7784	7784	0.177	43967	41.67	0.00340
21	0	22	15	75	0.0142	1764	250	0	2238	2238	0.130	17201	16.30	0.00354
22	0	22	15	97	0.0184	1764	250	0	2238	2238	0.130	17201	16.30	0.00367
23	0	22	15	119	0.0225	1764	250	0	2238	2238	0.130	17201	16.30	0.00381
24	0	22	15	141	0.0267	1764	250	0	2238	2238	0.130	17201	16.30	0.00394
25	0	22	15	163	0.0309	1764	250	0	2238	2238	0.130	17201	16.30	0.00408
26	0	22	15	185	0.0350	1764	250	0	2238	2238	0.130	17201	16.30	0.00421
27	0	22	15	207	0.0392	1764	250	0	2238	2238	0.130	17201	16.30	0.00435
28	0	22	15	229	0.0434	1764	250	0	2238	2238	0.130	17201	16.30	0.00448

off at a stoplight or stop sign or cruising at low speeds. With a hybrid electric car, we will use the electric motor when the internal combustion engine is operating at low thermal efficiency. The internal combustion engine will power the car and charge the battery when it is operating at a higher thermal efficiency. The hybrid vehicle requires an internal combustion engine and an electric motor. It is more complex than gasoline vehicles, but there is a potential for significant improvement in fuel economy. The improvement in fuel economy comes from being able to turn the gasoline engine off when the vehicle is stopped, and by allowing the gasoline engine to operate at a

FIGURE 11.5   Instantaneous fuel economy.

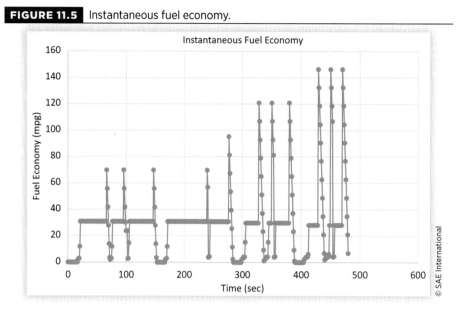

© SAE International

higher thermal efficiency when it is on. The electric motor is used to power the vehicle when taking off from stoplights, or other instances when the gasoline engine would be operating at low thermal efficiency. Another benefit to the hybrid vehicle is that we can recover some of the braking energy with regenerative braking. The electric motor in the car is used as a generator, and the wheels and drive train power the generator to produce the electric energy and charge the batteries.

All the energy for powering the car comes from the internal combustion engine because it charges the batteries that provide the power for the electric motor. If the battery is large enough to provide power for a few minutes, the electric motor can also supplement the power of the internal combustion engine when the car is climbing a steep hill or accelerating. This gives the car more power than can be provided by the internal combustion engine alone. Since the electric motor can supplement the power of the internal combustion engine, we can use a smaller internal combustion engine than would be required without the electric motor. Using a smaller engine will allow us to further increase the overall thermal efficiency of the engine. In Chapter 10, there is a section illustrating how using a smaller engine will allow the vehicle to get better fuel economy.

The control logic of when to use the gasoline engine and when to use the electric motor is complex. The car companies are still working out how to optimize the hybrid electric car. The battery should be large enough to power the car for a few minutes. Larger batteries provide more flexibility in deciding when the gasoline engine needs to be started. Optimizing the size of the battery for the car is also a difficult problem that is not fully understood at this point in time. The hybrid electric cars are complex and expensive, but they offer the potential of a large improvement in urban fuel economy compared to a gasoline-powered car or truck. The improvement in fuel economy comes without any sacrifice in the performance of the vehicle. Hybrid vehicles are filled with gas like a gasoline-powered vehicle, which makes them more convenient than electric vehicles when making a long trip. Electric vehicles have limited range and it takes a while to charge the batteries.

The gasoline engine will need to run continuously on the highway to provide the necessary power. The electric motor can assist when the gasoline engine needs to produce more power, and this allows a smaller gasoline engine to be used than would be required for a standard gasoline-powered car. A hybrid electric car will deliver an improvement in the highway fuel economy, but the improvement will not be as large as the improvement in urban fuel economy.

Example 11.8: Consider what would happen if we make a hybrid version of a pickup truck (modeled after the F150). The weight is 5500 lb, Crr = 0.01, Drag Area = 2.77 m^2 and the aero drag coefficient is 0.5. Assume the transmission and drive are 90% efficient in transmitting power from the engine to the wheels. The truck will use the electric motor in conjunction with the gas motor, so a smaller gas engine can be used. Assume the gas engine is rated at 150 hp. Using a smaller gas engine improves thermal efficiency.

Discussion: We will start with the spreadsheet developed for the truck (Table 11.12). The idle power can be set to zero because the gas engine will be turned off when the truck stops. The rest of the modifications to the spreadsheet depend on how the control system is programmed to decide between using the gas engine and the electric motor. Ideally, the electric motor will be used when the truck is traveling at low speed or a low power requirement. The gas motor will be started to charge the batteries and power the car at the same time, which will allow the gas engine to operate at a higher thermal efficiency on average. The amount of time the truck can run on electric power depends on the size of the battery system.

I hope the reader can appreciate the complexity in optimizing the performance and fuel economy of hybrid vehicles. We spent decades optimizing gasoline and diesel vehicles. Having two sources of power (fuel and electric) and a finite-sized battery makes optimizing the hybrid vehicle an order of magnitude more complex. We want to deliver adequate power and maximize fuel economy. It will require decades of research and experimentation to fully optimize the hybrid vehicle. Simulations will be part of the development effort. The models will be modified according to the strategy used in deciding when to use the gas engine or the electric motor. The solution below is one possibility in how the hybrid vehicle could be used in urban traffic.

Solution: For this model assume that the control logic will be set so that the gas engine will operate at a minimum thermal efficiency of 15%. Assume that 30% of the braking energy will be recovered and put into the batteries. Assume that the battery is large enough to allow the vehicle to accomplish these two things. Developing the software and control system to accomplish this would be difficult but assume that it is possible.

The previous model for the truck can be modified to generate the model for the hybrid electric vehicle. The changes are listed below.

1. The total power column is rolling + aerodynamic + kinetic. It should not be divided by the efficiency of the gas engine because it will not be all gas-powered. Portions will be electrically powered.

2. The "cleaned" power column should have a logic statement saying that if the total power is positive use the total power divided by the drive efficiency. If the total power is negative it should be multiplied by the regenerative braking efficiency.

$$\text{If}\left(\text{Total} > 0, \frac{\text{Total}}{\text{drive eff}}, \text{Total} * \text{regen eff}\right) \tag{11.24}$$

3. The thermal efficiency needs to be separated into two columns because of the way Excel works. We need to use two logic statements to work it out. For the first column if the cleaning power is negative or zero set the thermal efficiency to zero. Otherwise, set it to the logarithmic formula used for gasoline engines. If you leave out this step the negative powers will cause problems in the logarithmic equation.

$$\text{If}\left(P > 0, (0.0384)\ln\left[\frac{124\,P}{375} + 25.9\right] - 0.1249, 0\right) \tag{11.25}$$

4. A column is added for minimum efficiency. In the example, the minimum efficiency is set to 15%. It is assumed in this example that the gas engine will only be turned on when it can operate at an efficiency of at least 15%. It is assumed the electric motor will need to power the car in all cases when the gas engine would need to operate at less than 15% thermal efficiency. The logic statement here is that if the thermal efficiency in the previous column is less than the minimum, then the minimum will be placed in this column. If the thermal efficiency is greater than the minimum, then the value calculated by the logarithmic equation will be used.

If the parameters in the spreadsheet are changed to reflect a hybrid F-150 pickup (W = 5500 lb, Crr = 0.01, A = 2.77 m^2, Cd = 0.5, P$_{max}$ = 245 hp, drive eff = 0.90) the truck would get 26.3 mpg in city driving. This would be an excellent fuel economy for such a large vehicle. The spreadsheet developed for this example is shown in Example 11.9 below.

Example 11.9: The Toyota Prius is a hybrid car. Weight is 3500 lb, Crr = 0.009, drag area = 1.82 m^2 and the aero drag coefficient is 0.31. The engine is rated at 121 hp and the drive efficiency is about 90%. Develop the spreadsheet and estimate the fuel economy of the Toyota Prius in city driving.

Solution: The model yields a fuel economy of 47.5 mpg for the Toyota Prius, which is very near the EPA city rating for the car. The first 30 s of the spreadsheet model for the Toyota Prius is shown in **Table 11.13** below.

TABLE 11.13 Urban driving model for a hybrid electric car.

Car Weight	3500	lb		Pmax	121	hp			Fuel Energy	121000	BTU/ gal				
Crr	0.009			eff	0.9									min eff	0.15
Area	1.82	m²		idle	0	W			Econ	47.5	mpg			regen eff	0.3
Cd	0.31														
s	ft/s²	ft/s	mph	feet	Miles	W	W	W	W	W	Thermal	min	W	BTU/s	gallons
Time	Acc	Speed	Speed	Dist	Dist	Rolling	Aero	Kinetic	Total	Clean	Effi	eff	Fuel	Fuel	fuel
0	0	0	0	0	0	0	0	0	0	0	0.000	0.150	0	0.00	0
1	0	0	0	0	0	0	0	0	0	0	0.000	0.150	0	0.00	0
2	0	0	0	0	0	0	0	0	0	0	0.000	0.150	0	0.00	0
3	0	0	0	0	0	0	0	0	0	0	0.000	0.150	0	0.00	0
4	0	0	0	0	0	0	0	0	0	0	0.000	0.150	0	0.00	0
5	0	0	0	0	0	0	0	0	0	0	0.000	0.150	0	0.00	0
6	0	0	0	0	0	0	0	0	0	0	0.000	0.150	0	0.00	0
7	0	0	0	0	0	0	0	0	0	0	0.000	0.150	0	0.00	0
8	0	0	0	0	0	0	0	0	0	0	0.000	0.150	0	0.00	0
9	0	0	0	0	0	0	0	0	0	0	0.000	0.150	0	0.00	0
10	0	0	0	0	0	0	0	0	0	0	0.000	0.150	0	0.00	0
11	0	0	0	0	0	0	0	0	0	0	0.000	0.150	0	0.00	0
12	0	0	0	0	0	0	0	0	0	0	0.000	0.150	0	0.00	0
13	0	0	0	0	0	0	0	0	0	0	0.000	0.150	0	0.00	0
14	0	0	0	0	0	0	0	0	0	0	0.000	0.150	0	0.00	0
15	0	0	0	0	0	0	0	0	0	0	0.000	0.150	0	0.00	0
16	7	0	0	0	0	0	0	0	0	0	0.000	0.150	0	0.00	0
17	7	7	4.77	3.5	0.001	306	3.3	3620	3929	4366	0.198	0.198	22036	20.89	0.0002
18	7	14	9.55	14	0.003	627	26.3	10859	11512	12791	0.239	0.239	53463	50.68	0.0006
19	1	21	14.32	31.5	0.006	961	88.7	18098	19148	21275	0.259	0.259	82223	77.94	0.0012
20	0	22	15	53	0.010	1010	102.0	3176	4289	4765	0.201	0.201	23654	22.42	0.0014
21	0	22	15	75	0.014	1010	102.0	0	1112	1236	0.150	0.150	8228	7.80	0.0015
22	0	22	15	97	0.018	1010	102.0	0	1112	1236	0.150	0.150	8228	7.80	0.0015
23	0	22	15	119	0.023	1010	102.0	0	1112	1236	0.150	0.150	8228	7.80	0.0016
24	0	22	15	141	0.027	1010	102.0	0	1112	1236	0.150	0.150	8228	7.80	0.0017
25	0	22	15	163	0.031	1010	102.0	0	1112	1236	0.150	0.150	8228	7.80	0.0017
26	0	22	15	185	0.035	1010	102.0	0	1112	1236	0.150	0.150	8228	7.80	0.0018
27	0	22	15	207	0.039	1010	102.0	0	1112	1236	0.150	0.150	8228	7.80	0.0019
28	0	22	15	229	0.043	1010	102.0	0	1112	1236	0.150	0.150	8228	7.80	0.0019

At this point in time, we are working to develop electric cars that have a good driving range and a fast charging time. The vision is that if all the electricity is produced by renewable energy (wind, solar, hydro), the electric vehicles will have zero emissions. We have made a lot of progress and will probably be able to develop small and medium-sized cars that have an acceptable range. The larger vehicles, especially large trucks will require a lot of batteries, and it may not be possible to provide an adequate range with lithium-ion batteries. Charging time is a more difficult problem to solve than range. At this point in time, it seems unlikely to me that we will be able to develop a battery system that can be charged in 5 or 10 min, which is typical for how long it takes to fill a vehicle with gasoline. The hybrid vehicle or plug-in hybrid vehicle may be the best overall compromise.

Homework

1. Assume that we do careful fuel economy testing on a light truck that weighs 5400 lb, Crr = 0.012, the aerodynamic drag area is 2.8 square meters, and Cd = 0.47. Assume the truck is burning 10% ethanol–gasoline blend that has 121,000 BTU/gallon.

   a. Cruising on the level pavement at a steady 30 mph the truck gets 26.1 mpg. What is the efficiency of the engine/drive system at 30 mph? [Efficiency of the engine/drive system is the product of the thermal efficiency of the engine and the efficiency of the drive system in getting the power to the wheels.] [Answer 15.58%]

   b. Cruising on the level pavement at a steady 60 mph the truck gets 21.9 mpg. What is the efficiency of the engine/drive system at 60 mph? [Answer: 26.02%]

   c. Assume that the efficiency of the engine/drive system is zero at zero mph. Plot the efficiency of the engine drive system vs speed over the range of 0 mph to 60 mph using the data points in a and b above and the point (0,0). Draw a trend line through the data points and have excel calculate the slope of the trend line. Please use only the three data points for the plot, 0% efficiency at 0 mph, and the efficiencies you get for 30 mph and 60 mph.

2. Model an electric car driving on the SAE J1082 suburban cycle. Parameters for the car are W = 4400 lb, Crr = 0.009, A = 1.9 m², Cd = 0.31, motor/drive efficiency = 88%, battery capacity = 60,000 watt-hrs.

   a. Assume that there is no regenerative braking on the car and calculate the number of watt-h per mile energy consumption, and the range of the vehicle for suburban driving. [Answer: 270 watt-h/mile, 222 miles]

   b. Assume that regenerative braking is added and that 30% of the braking energy will be recovered and put back into the batteries. Calculate the number of watt-h per mile energy consumption, and the range of the vehicle for suburban driving. [Answer 245 watt-h/mile, 245 miles]

   c. Evaluate the impact of adding regenerative braking in increasing the range of the car in suburban driving. What is the increase in the range of the car from adding regenerative braking?

   d. Suppose that by using exotic materials we can reduce the weight of the vehicle 500 lb. What is the increase in the range of the car from reducing the weight by 500 lb? [Compare to the original case with no regenerative braking and only change the weight of the vehicle.]

e. Suppose that we can make considerably more luggage room in the hatchback region by squaring off the back of the car like an SUV rather than having it sloped and more aerodynamic. This will increase the drag coefficient to Cd = 0.4. What is the decrease in the range of the car from this modification? [Compare to the original case with no regenerative braking and only change the drag coefficient of the vehicle.]

f. Suppose we can make the car more pleasant to drive by using wider tires that operate at a lower pressure. This will increase Crr to a value of 0.014. What is the decrease in the range of the car from this modification? [Compare to the original case with no regenerative braking and only change the rolling coefficient of the vehicle.]

3. Model a small gasoline car driving on the SAE J1082 suburban cycle. Parameters for the car are W = 2800 lb, Crr = 0.010, A = 1.9 m², Cd = 0.32, drive efficiency = 90%. The power rating for the engine is 130 hp and burns the 10% ethanol–gasoline blend that has 121,000 BTU per gallon.

a. Use the method developed in class and calculate the mpg fuel economy of the car driving in a suburban environment. If the car has a 13.2-gal tank, what is the range of the vehicle? [Answer 40.2 mpg]

b. Suppose that by using exotic materials we can reduce the weight of the vehicle 500 lb. What is the increase in mpg of the car from reducing the weight by 500lb ? [Compare to the original case and only change the weight of the vehicle.]

c. Suppose that we can make considerably more luggage room in the hatchback region by squaring off the back of the car like an SUV rather than having it sloped and more aerodynamic. This will increase the drag coefficient to Cd = 0.42. What is the decrease in mpg of the car from this modification? [Compare to the original case and only change the drag coefficient.]

d. Suppose we can make the car more pleasant to drive by using wider tires that operate at a lower pressure. This will increase Crr to a value of 0.014. What is the decrease in mpg of the car from this modification? [Compare to the original case and only change the rolling coefficient.]

4. [Repeat #3 for the urban cycle] Model a small gasoline car driving on the SAE J1082 urban cycle. Parameters for the car are W = 2800 lb, Crr = 0.010, A = 1.9 m², Cd = 0.32, drive efficiency = 90%. The power rating for the engine is 130 hp and burns the 10% ethanol–gasoline blend that has 121,000 BTU per gallon.

a. Use the method developed in class and calculate the mpg fuel economy of the car driving in an urban environment. If the car has a 13.2-gal tank, what is the range of the vehicle? [Answer: 39.5 mpg]

b. Suppose that by using exotic materials we can reduce the weight of the vehicle 500 lb. What is the increase in mpg of the car from reducing the weight by 500 lb? [Compare to the original case and only change the weight of the vehicle.]

   c. Suppose that we can make considerably more luggage room in the hatchback region by squaring off the back of the car like an SUV rather than having it sloped and more aerodynamic. This will increase the drag coefficient to Cd = 0.42. What is the decrease in mpg of the car from this modification? [Compare to the original case and only change the drag coefficient of the vehicle.]

   d. Suppose we can make the car more pleasant to drive by using wider tires that operate at a lower pressure. This will increase Crr to a value of 0.014. What is the decrease in mpg of the car from this modification? [Compare to the original case and only change the rolling coefficient of the vehicle.]

5. [Repeat #3 for the 70-mph highway cycle] Model a small gasoline car driving on the SAE J1082 highway cycle. Assume that the thermal efficiency of the engine is a constant 30% for this case. Parameters for the car are W = 2800 lb, Crr = 0.010, A = 1.9 m², Cd = 0.32, drive efficiency = 90%. The power rating for the engine is 130 hp and burns the 10% ethanol–gasoline blend that has 121,000 BTU per gallon.

   a. Use the method developed in class and calculate the mpg fuel economy of the car driving on the highway. If the car has a 13.2-gal tank, what is the range of the vehicle? [Answer 41.0 mpg]

   b. Suppose that by using exotic materials we can reduce the weight of the vehicle 500 lb. What is the increase in mpg of the car from reducing the weight by 500 lb? [Compare to the original case and only change the weight of the vehicle.]

   c. Suppose that we can make considerably more luggage room in the hatchback region by squaring off the back of the car like an SUV rather than having it sloped and more aerodynamic. This will increase the drag coefficient to Cd = 0.42. What is the decrease in mpg of the car from this modification? [Compare to the original case and only change the drag coefficient of the vehicle.]

   d. Suppose we can make the car more pleasant to drive by using wider tires that operate at a lower pressure. This will increase Crr to a value of 0.014. What is the decrease in mpg of the car from this modification? [Compare to the original case and only change the rolling coefficient of the vehicle.]

6. Model a small hybrid gasoline car driving on the SAE J1082 urban cycle. Parameters for the car are W = 3100 lb, Crr = 0.010, A = 1.9 m², Cd = 0.32, drive efficiency = 90%. The power rating for the engine is 130 hp and burns the 10% ethanol–gasoline blend that has 121,000 BTU per gallon. Assume that the electric motor and battery are designed such that the engine will always operate at 15% thermal efficiency or higher. Also, assume the car has regenerative braking and that 30% of the braking energy will be recovered.

   a. Use the method developed in the book and calculate the mpg fuel economy of the hybrid car driving in an urban environment. If the car has a 13.2-gal tank, what is the range of the vehicle? [48.4 mpg]

b. Suppose that by using exotic materials we can reduce the weight of the vehicle 500 lb. What is the increase in mpg of the car from reducing the weight by 500 lb? [Compare to the original case and only change the weight of the vehicle.]

c. Suppose that we can make considerably more luggage room in the hatchback region by squaring off the back of the car like an SUV rather than having it sloped and more aerodynamic. This will increase the drag coefficient to Cd = 0.42. What is the decrease in mpg of the car from this modification? [Compare to the original case and only change the drag coefficient of the vehicle.]

d. Suppose we can make the car more pleasant to drive by using wider tires that operate at a lower pressure. This will increase Crr to a value of 0.014. What is the decrease in mpg of the car from this modification? [Compare to the original case and only change the rolling coefficient of the vehicle.]

11.5 **Effect of Hills on Fuel Economy**

Grades on interstate highways seldom exceed 4% for most of the country. In mountainous areas, it is common to have 6% or 7% grades on the steeper parts of the mountains. With a 4% grade, the road rises 4 feet for every 100 ft traveled, or the sine of the angle is 4/100, which is 2.29 degrees. The angle of the road seems a lot higher than this, but the grades on the roads are very small angles. The power to overcome gravity is equal to the weight of the vehicle multiplied by the upward component of velocity.

Example: Suppose an 80,000-lb truck climbs a 4% grade at 70 mph. What is the horsepower required to overcome gravity? For this example, neglect rolling resistance and aerodynamics and only calculating the portion of horsepower required to climb the hill.

Solution: The upward component of velocity is 70 mph multiplied by 4%. The power to overcome gravity is calculated as follows:

$$\text{Gravity Power} = (80,000\,\text{lb})(70\,\text{mph})(0.04)\left(\frac{5280}{3600}\right) = 328,533\,\text{ft-lb/s} \quad (11.26)$$

Since 1 hp is equal to 550 ft-lb/s, the truck requires 597 hp to overcome the gravity loading. Similar calculations show that 896 hp is required for a 6% grade and 1045 hp is required for a 7% grade. The truck will require an additional 250 hp to overcome rolling resistance and aerodynamics. These are large amounts of horsepower, even for an 18-wheeler truck. The 18-wheelers typically have 400 to 600 hp available to push the truck up the hill. No fully loaded (80,000 lb) truck can climb a 4% grade at 70 mph. If the truck is unloaded so it weighs only 40,000 lb then it is feasible for the truck to climb a 4% grade at 70 mph. Hill climbing requires a substantial amount of horsepower.

Example 11.10: A medium 6-cylinder pickup truck is pulling a boat traveling at 80 mph as it approaches a hill with a 7% grade. The weight of the truck and contents is 5000 lbs and the weight of the boat is 3000 lbs. The rolling resistance coefficient for the tires is 0.011. The aerodynamic drag coefficient is 0.85 and the drag area is 3.8 square meters. The truck provides a maximum of 150 hp to the wheels.

a. Calculate the steady-state speed of the truck up the hill.

b. Develop a spreadsheet model to show how the velocity decreases as the truck starts up the hill.

Solution: Part a can be solved analytically. If we work in watts the gravity, rolling and aerodynamic powers can be expressed as a function of speed and we can solve for the steady-state speed.

$$\text{Gravity Power} = (8000)(V)\left(\frac{5280}{3600}\right)\left(\frac{7}{100}\right)\left(\frac{745.7}{550}\right) \tag{11.27}$$

$$\text{Rolling Power} = (1.989)(8000)(V)(0.011)\left(1 + \frac{V}{200}\right) \tag{11.28}$$

$$\text{Aero Power} = (0.05357)(V^3)(3.8)(0.85) \tag{11.29}$$

$$\text{Gravity} + \text{Rolling} + \text{Aero} = (150)(745.7) \tag{11.30}$$

The steady-state speed is calculated by solving Eq. 11.30 as 58.1 mph. The truck and trailer would be traveling 80 mph when it got to the hill and would gradually slow down to 58.1 mph. It would be able to maintain the 58.1 mph the rest of the way up the hill.

For part b of the problem, we will use a spreadsheet approach and develop an approximate model. Kinetic power is the product of mass, velocity, and acceleration, and acceleration is the derivative of velocity with time. In principle, it is possible to set up a first-order differential equation and generate an analytical solution. It is much easier and more straightforward to use the spreadsheet approach to generate an approximate solution.

For this example problem, I chose to step through the process in 0.5-s intervals. The truck starts at 80 mph and from that number, we calculate the power to overcome gravity, rolling, and aerodynamic resistance. The engine can only provide 150 hp, and at 80 mph that will not be enough for the truck to maintain speed. The difference in power is made up by a reduction in kinetic energy over the 0.5 s time step. The energy balance at each

TABLE 11.14 First three steps of a truck pulling boat uphill.

Time (s)	Speed (mph)	Gravity (W)	Rolling (W)	Aero (W)	Engine (W)	Kinetic (J)	$(V_2)^2$ $(mph)^2$
0	80	89,088	19,600	88,576	111,855	42,705	6282.13
0.5	79.26	88,264	19,367	86,141	111,855	40,959	6169.1
1.0	78.54	87,466	19,143	83,826	111,855	32,290	6060.7

time step will require a small change in velocity of the truck, which will allow us to find the velocity as a function of time for the truck. Parameters for the vehicle are W = 8000 lb, Crr = 0.011, A = 3.8 m², Cd = 0.85, Power = 150 hp, Grade = 7%, and Initial Speed = 80 mph. The first three steps of the spreadsheet are shown in the **Table 11.14**:

Gravity, rolling, and aero power are calculated using Eqs. 11.27, 11.28, and 11.29 using the speed in column 2 of the table. In this model it is assumed that the engine and drive can provide 150 hp to the wheels, which is 111,855 W. The engine is not capable of providing enough power to overcome the sum of gravity, rolling and aerodynamic loads, so the difference must be made up with a reduction in kinetic energy. The kinetic energy is the power difference multiplied by 0.5 s time step in Eq. 11.31.

$$\text{Kinetic Joules} = \left(\text{Gravity} + \text{Rolling} + \text{Aero-Engine}\right)\left(0.5\,\text{s}\right) \qquad (11.31)$$

If V_1 is the velocity at the beginning of the time step in column 2 and V_2 is the velocity at the end of the time step the change in kinetic energy is ½ m $(V_1^2 - V_2^2)$. The speeds are in mph and the weight is given in pounds so there are unit conversions, but the equation can be solved for V_2^2 as shown in Eq. 11.32 below.

$$V_2^2 = V_1^2 - \frac{(22.0811)(\text{Kinetic Joules})}{\text{Weight}} \qquad (11.32)$$

Take the square root of V_2^2 in column 8 to get the value for V_1 in the next row in column 2. The speeds in column 2 are the speed of the truck and trailer at the beginning of each time step. Energy is balanced by the truck and trailer slowing down a little each time step. The reader should be able to duplicate the numbers in the table. Excel results for the first 5 s are shown in **Table 11.15**. The first 70 s are graphed in Figure 11.5.

In the model the speed of the truck and trailer decreases gradually to the 58.1 mph steady-state speed calculated in part a of the problem. Reality is more complex than is illustrated in the model because the engine will not produce a constant 150 hp over the range of speeds. The engine power will be approximately constant, but the engine RPM will depend on the vehicle speed and engine power varies with RPM. The transmission may downshift, which further complicates the issue. But all things considered, this is a good model to illustrate quantitatively how hill climbing impacts the energy consumption and speed of the vehicle (**Figure 11.6**).

TABLE 11.15 Spreadsheet of a truck pulling boat uphill.

Vehicle Weight	8000	lb					
Roll Resistance	0.011						
Drag Coeff	0.85						
Drag Area	3.8	sq. m					
Engine Power	150	hp					
Hill Grade	0.07	grade					
Initial Speed	80	mph					
	Speed	Gravity	Rolling	Aero	Engine	Kinetic	V2
Time Steps	mph	W	W	W	W	J	squared
0	80	89086.29	19603.58	88591.92	111855	42713.4	6282.105
0.5	79.25973	88261.95	19370.84	86155.29	111855	40966.54	6169.032
1	78.54318	87464.01	19146.46	83839.69	111855	39297.59	6060.565
1.5	77.84963	86691.69	18930.14	81638.28	111855	37702.56	5956.501
2	77.17837	85944.19	18721.58	79544.65	111855	36177.71	5856.645
2.5	76.52872	85220.75	18520.48	77552.82	111855	34719.52	5760.815
3	75.90003	84520.66	18326.57	75657.16	111855	33324.69	5668.834
3.5	75.29166	83843.19	18139.59	73852.43	111855	31990.1	5580.537
4	74.70299	83187.66	17959.28	72133.69	111855	30712.81	5495.765
4.5	74.13343	82553.41	17785.4	70496.32	111855	29490.06	5414.369
5	73.58239	81939.78	17617.71	68935.97	111855	28319.24	5336.204

© SAE International

FIGURE 11.6 Graph of a truck pulling boat uphill.

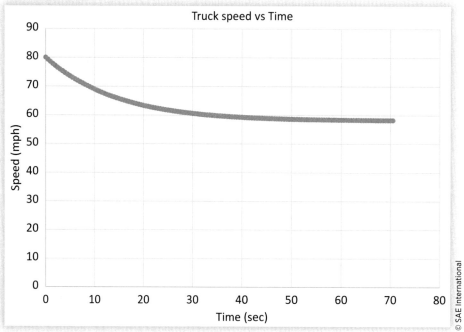

© SAE International

11.6 Development of a Highway Model

To develop a model of vehicles driving on typical highways we could survey the route and get accurate survey data. This would take a long time and be expensive. Another way is to drive the route in a vehicle with a Global Positioning System (GPS) connected to a computer and have the GPS collect latitude, longitude, and altitude along the route. The reader should imagine a huge spreadsheet with the first three columns of the spreadsheet being the latitude, longitude, and altitude of points along the highway route. This information can be used to develop a digital model of the highway, and it can be done quickly with relatively inexpensive equipment.

The earth is approximately a spherical polar coordinate system. Assume R is the radius of the earth, θ_{Lat} is the latitude, θ_{Long} is the longitude, and h is the altitude above sea level. The average earth radius is 3959 miles or 6371 kilometers. The radius is smaller at the poles (6353 km) and larger at the equator (6384 km). We are interested in areas in latitudes such as the United States, Europe, Asia, and Australia, so the average radius is the best number to use. An (x,y,z) cartesian coordinate system can be imposed as follows:

$$r = R + h \tag{11.33}$$

$$x = r \cos(\theta_{Lat}) \cos(\theta_{Long}) \tag{11.34}$$

$$y = r \cos(\theta_{Lat}) \sin(\theta_{Long}) \tag{11.35}$$

$$z = r \sin(\theta_{Lat}) \tag{11.36}$$

The distance between two points in the model is given by the distance formula:

$$\text{Distance} = \left[(x_2 - x_1)^2 + (y_2 - y_1)^2 + (z_2 - z_1)^2 \right] \tag{11.37}$$

The average grade along the distance is the change in height divided by the distance:

$$\text{Grade} = \frac{h_2 - h_1}{\text{Distance}} \tag{11.38}$$

Table 11.16 below shows three hypothetical data points. To be sure and understand the process the reader should be able to duplicate the numbers in the table.

It is not possible to calculate the distance or slope for the first row in the table because two data points are required to make the calculations. There will be a noise factor in the GPS data, and it may be necessary to do some smoothing to obtain a reasonable digital approximation of the highway. The file used in this book is a model of highway 63 heading

TABLE 11.16 Converting latitude, longitude, and altitude into a digital model of the highway.

Latitude	Longitude	Altitude	Radius	x-location	y-location	z-location	Distance	Slope
Degrees	Degrees	m	m	m	m	m	m	Grade
37.95143	−91.77127	342.5	6371342.5	−155289.98	−5021609.3	3918332.646	0	
37.95138	−91.77122	342.2	6371342.2	−155285.696	−5021612.6	3918328.078	7.0870	−0.04233
37.95133	−91.77125	342.1	6371342.1	−155288.429	−5021615.8	3918323.632	13.2388	−0.01626

south out of Rolla, Missouri. It is a road that goes through the Ozark hills and is hilly terrain. The model can be used to illustrate how hilly terrain impacts the fuel economy and energy consumption of the vehicle.

The model has the distance along the route and the slope of the road in percent grade at each point along the route. It was developed from GPS data as illustrated in Table 11.16. Assuming the weight of the vehicle is in lb and the speed in mph the gravity power in W is given by Eq. 11.39.

$$\text{Gravity Power} = (1.989) \, W \, V \, (\text{Grade}) \, W \qquad (11.39)$$

Table **11.17** illustrates how the model is started. The parameters for this example are W = 3800 lb, Crr = 0.011, A = 1.9 m², Cd = 0.32, and V = 50 mph.

There are 8000 lines of data for the model that gives the distance along the route and the slope of the road in percent grade. The total distance is 175 miles. The maximum grade is less than 4% on this route. A negative grade means it is downhill and a positive grade is uphill. For example, we ill model the vehicle going at a constant speed of 50 mph. The reader should imagine a table 8000 lines in length where the number for the distance along the route and slope are known, and the rest of the table is blank. We would start at time zero, so the first row under the two-time columns are zero.

The time in hours in column 2 is the distance in column 3 divided by the speed. The time in seconds for column 1 is the time in hours in column 2 multiplied by 3600. Rolling, aero, and gravity power are computed from Eq. 11.26, Eq. 11.27, and Eq. 11.28 and then totaled in column 8. The first 8 columns in the spreadsheet are the same whether it is an electric car or a gasoline-powered car. From this information, the reader should be able to duplicate **Table 11.18**, which is the same as Table 11.17 with the blanks filled in.

For electric vehicles, the total power needs to be "cleaned." If the total power is positive it must go through the electric motor and drive system which is typically 85% efficient. If the total power is negative, regenerative braking will be able to recover about 30% of the energy. The energy used or added in each step is subtracted or added to the battery (**Table 11.19**).

TABLE 11.17 Starting the highway simulation model.

Time (s)	Time (h)	Distance (Mi)	Slope Grade (%)	Rolling (W)	Aero (W)	Gravity (W)	Total (W)
0	0	0	−3.46				
		0.01352	−2.27				
		0.02844	−1.73				

© SAE International

TABLE 11.18 Completion of Table 11.17.

Time (s)	Time (h)	Distance (Mi)	Slope Grade (%)	Rolling (W)	Aero (W)	Gravity (W)	Total (W)
0	0	0	−3.46	5196.26	4071.32	−13,076	−1142.5
0.97344	.000270	0.01352	−2.27	5196.26	4071.32	−8579	810
2.04768	.000569	0.02844	−1.73	5196.26	4071.32	−6538	3211

© SAE International

TABLE 11.19 Electric car on highway example.

Car Weight	3800			Battery	30,000		Based on the first 50 miles		
Crr	0.011	lb		eff	0.85	watt-h	W-h/mi	225.1648	
Area	1.9	m^2		Speed	50	mph	range	133.2358	miles
Cd	0.32			regen	0.3				
Time	Time	Distance the Route	Inclination the Road	Rolling	Aero	Hill	Total Power	Cleaned	Battery
(s)	(h)	(M)	(Grade %)	(W)	(W)	(W)	(W)	(W)	(W·h)
0.00	0.00000	0.000	−3.50	5196	4071	−13219	−3952	−1185	30000
0.97	0.00027	0.014	−3.27	5196	4071	−12355	−3087	−926	30000
2.05	0.00057	0.028	−2.31	5196	4071	−8715	552	650	30000
3.22	0.00089	0.045	−1.77	5196	4071	−6674	2594	3052	29999
4.43	0.00123	0.062	−1.43	5196	4071	−5418	3849	4528	29998
5.66	0.00157	0.079	−1.15	5196	4071	−4357	4911	5777	29996
6.90	0.00192	0.096	−0.48	5196	4071	−1822	7446	8760	29994
8.15	0.00226	0.113	−0.17	5196	4071	−638	8630	10153	29991
9.43	0.00262	0.131	0.29	5196	4071	1083	10351	12177	29987
10.74	0.00298	0.149	0.66	5196	4071	2506	11774	13852	29982
12.11	0.00336	0.168	1.37	5196	4071	5167	14434	16982	29976
13.47	0.00374	0.187	1.27	5196	4071	4811	14079	16564	29970

The parameters used in the table are correct for the 2017 Nissan Leaf. In this example, I chose to use the first 50 miles to estimate the W-h/mi energy efficiency and range of the vehicle. There is nothing magical about the 50-mi choice, as long as you choose a distance of at least 50 miles. There are enough hills in the first 50 miles that the average efficiency of the vehicle has been established. Choosing a longer distance will yield a slight, but the insignificant difference in the results. The model shows that the car would use 225 W-h of energy per mile traveled and have a range of 133 miles. If the electricity cost is $0.12 per kW-h, the energy cost for the car would be $0.02 per mi. A small gasoline car that gets 40 mpg on the highway with a fuel cost of $2.00 per gal would cost $0.05 per mile. The energy cost per mile for the electric car is very low compared to gas cars, but the range of 133 mi would be problematic for most people when traveling on the highway. If the speed is increased to 70 mph the range decreases to 92 miles. That's enough for people to use in daily commutes but it is not enough for making a long trip on the highway.

The next step is to model a gasoline-powered car driving along the route. For this example, it was assumed that the engine has a constant thermal efficiency of 28%. It is possible to use the logarithmic expression for engine thermal efficiency (Eq. 11.21), but for highway driving this equation underestimates the average thermal efficiency of the engine. Eq. 11.21 was developed by fitting a curve through the average thermal efficiency of the engine, which is typical for city driving. For highway driving the transmission will choose a higher gear and allow the engine to operate nearer the maximum thermal efficiency (**Table 11.20**).

For comparison, the parameters used in Table 11.20 are for a gasoline car that is the same size as the Nissan Leaf. The gas car would be lighter in weight but would have the

TABLE 11.20 Gasoline car on highway example.

Car Weight	2800	lb		Max Power	130	hp			Fuel Economy	55.9	mpg	
Crr	0.011			Idle Power	193.88	W						
Area	1.9	m^2		Drive eff	0.9				Engine Eff	0.28		
Cd	0.32			Speed	50	mph						
		Distance the Route	Inclination the Road						Engine Efficiency	Fuel Power	Fuel Power	Fuel Used
Time	Time			Rolling	Aero	Hill	Total Power	Cleaned				
(s)	(h)	(M)	(Grade %)	(W)	(W)	(W)	(W)	(W)		(W)	(BTU/sec)	(gal)
0.00	0.00000	0.000	−3.50	3829	4071	−9740	−1840	194	0.28	692	0.66	0.00000
0.97	0.00027	0.014	−3.27	3829	4071	−9103	−1203	194	0.28	692	0.66	0.00001
2.05	0.00057	0.028	−2.31	3829	4071	−6422	1478	1643	0.28	5866	5.56	0.00003
3.22	0.00089	0.045	−1.77	3829	4071	−4917	2983	3314	0.28	11836	11.22	0.00011
4.43	0.00123	0.062	−1.43	3829	4071	−3993	3908	4342	0.28	15506	14.70	0.00024
5.66	0.00157	0.079	−1.15	3829	4071	−3210	4690	5211	0.28	18610	17.64	0.00041
6.90	0.00192	0.096	−0.48	3829	4071	−1342	6558	7286	0.28	26023	24.67	0.00063
8.15	0.00226	0.113	−0.17	3829	4071	−470	7430	8256	0.28	29485	27.95	0.00090
9.43	0.00262	0.131	0.29	3829	4071	798	8698	9665	0.28	34516	32.72	0.00122
10.74	0.00298	0.149	0.66	3829	4071	1847	9747	10830	0.28	38678	36.66	0.00159
12.11	0.00336	0.168	1.37	3829	4071	3807	11707	13008	0.28	46457	44.04	0.00205
13.47	0.00374	0.187	1.27	3829	4071	3545	11445	12717	0.28	45418	43.05	0.00254

same Crr, area, and drag coefficient. The engine would produce about 130 hp and the average thermal efficiency on the highway would be about 28%. Driving at 50 mph this car would get 55.9 mpg, which is an excellent fuel economy, but 50 mph is a slow speed for the highway. If the speed is increased to 70 mph the fuel economy is 36.9 mpg, which is about right for a small car driving on the highway.

11.7 Summary of Unit 3

The purpose of Unit 3 was to introduce the reader to methods of using excel spreadsheets to simulate cars and trucks in urban (city), suburban, and highway driving. The models developed to estimate the energy consumption and fuel economy of the vehicles. The models are helpful in guiding the design and development of cars and trucks. When proposals are made to reduce weight, lower rolling resistance or improve the aerodynamics of the vehicle, the models help us assess quantitatively the improvement in energy efficiency or fuel economy that would result from the proposals. The models help us do a cost versus benefit analysis of proposed modifications so that we use our design and development resources wisely. Excel is very portable in the sense that almost everyone is comfortable using excel. Developing the models in excel makes it easier to pass the model to someone else. All the spreadsheet models developed for this book are available from the author.

In the first part of unit 3 we looked at the thermal efficiency of internal combustion engines and electric motors. I showed a method of starting with the engine map and developing an equation that relates the thermal efficiency of an internal combustion engine to the amount of power it produces. The model was developed for a Saturn 1.9L engine, but the method could be used to develop models for other engines, including diesel engines. The engine will have its best thermal efficiency when producing power at a low RPM, but the car will feel sluggish in city driving if the transmission keeps the engine operating at the RPM for maximum thermal efficiency. For city driving the transmission will choose a gear so that the engine is operating at an RPM such that the car is more pleasant to drive. For highway driving the transmission will tend to shift so that the engine is operating at a higher thermal efficiency.

In the middle part of Unit 3, I showed how to use the SAE J1082 specification to develop spreadsheet models for cars and trucks driving in urban (city), suburban, and highway driving. Models were developed for electric and gasoline-powered cars and trucks. The models allow us to quantitatively assess how the parameters (vehicle weight, rolling resistance of tires, aerodynamics, engine efficiency, and drive efficiency) impact the overall energy consumption and fuel economy of the vehicle.

In the last part of Unit 3, I showed how to develop a model for driving on a specific highway or race route. I was an advisor for the solar car project at Missouri S&T for many years, and to optimize the performance of the car on race routes, we developed models as described in this chapter. This was important for the solar cart project but less important for production vehicles. For production vehicles, the results modeling the vehicle traveling on level ground or on an actual highway are essentially the same. The models are to be used to estimate a difference in fuel consumption or electric energy consumption for a proposed modification to the vehicle. The difference will be almost the same if the vehicle is modeled traveling on level pavement or on a more realistic highway or street.

Homework

1. A small 4-cylinder pickup truck is pulling a boat traveling at 80 mph as it approaches a hill with a 6% grade. The weight of the truck and contents is 5800 lb, and the weight of the boat is 3200 lb. The rolling resistance coefficient for the tires is 0.012. The aerodynamic drag coefficient is 0.65 and the drag area is 3.8 square meters. The truck provides a maximum of 140 hp to the wheels.

   a. Calculate the steady-state speed of the truck up the hill. [Answer: 58.1 mph]

   b. Develop a spreadsheet model to show how the velocity decreases as the truck starts up the hill.

2. Model an electric car driving on the hilly terrain model. Parameters for the car are W = 4400 lb, Crr = 0.009, A = 1.9 m2, Cd = 0.31, motor/drive efficiency = 88%, battery capacity = 60,000 watt-hrs. The car drives at a steady 65 mph.

   Assume that there is no regenerative braking on the car and calculate the number of watt-h per mile energy consumption, and the range of the vehicle for driving in hilly terrain. [Answer: 278 watt-h/mile, 216 miles]

3. Model a small gasoline car driving on the hilly terrain model. Parameters for the car are W = 2800 lb, Crr = 0.010, A = 1.9 m², Cd = 0.32, drive efficiency = 90%. Assume that the thermal efficiency of the engine is 29%. The power rating for the engine is 130 hp and burns the 10% ethanol–gasoline blend that has 121,000 BTU per gallon. The car drives at a steady 65 mph.

Use the method developed in class and calculate the mpg fuel economy of the car driving in hilly terrain. If the car has a 13.2-gal tank, what is the range of the vehicle? [Answer: 43.7 mpg]

CHAPTER 12

Review of the Book (Course)

The book was developed to support a course about the energy efficiency of vehicles. At the end of the semester, the students take an exam over Unit 3, and then a week later take a comprehensive final exam. I end the course with a review starting with Unit 3 to prepare them for the last exam, and then after the last exam, I review for Units 1 and 2.

12.1 Review of Unit 3

The review is a series of example problems similar to what can be put on the exams. I can't ask them to develop spreadsheets for an exam, so I have them fill in the blanks in a table to illustrate that they know the formulas that would be used to build the spreadsheet. The focus of this review session is to review the types of problems I can put on the exams. **Figure 12.1** below will be used for some of the problems.

The engine map above is for a small diesel engine produced by Volkswagen to power some of their small cars. Assume that it takes 69.5 g of diesel fuel to provide 1 kW-h of heat energy.

FIGURE 12.1   Engine map for a diesel engine.

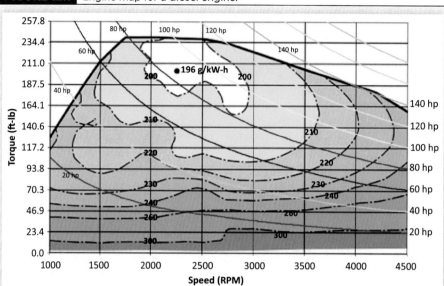

Example 12.1: The car in question is traveling 70 mph when the transmission is in high gear and the engine is turning 2500 RPM. When the car accelerates from 55 mph to 85 mph, the horsepower requirements from the engine are shown in **Table 12.1** below. The total hp in the table is the total power to overcome rolling, aero, and a 4ft/s² acceleration for the vehicle and assumes that the transmission and drive are 90% efficient in transmitting the power to the wheels. It is assumed the car is traveling on level ground. Fill in the table and plot the acceleration line on the engine map in Figure 12.1 above.

Solution: The first two columns in the table are given and you were asked to calculate the other two columns. The equations used are:

$$\text{Angular Velocity} = \frac{(2500)\,V}{70} \tag{12.1}$$

TABLE 12.1 Example 12.1

Speed (mph)	Total Horsepower	Angular Velocity (RPM)	Torque (ft-lb)
55	74.40		
60	82.84		
65	91.70		
70	101.02		
75	110.83		
80	121.17		
85	132.06		

RPM: Revolution per minute

TABLE 12.2 Completed Example 12.1.

Speed (mph)	Total Horsepower	Angular Velocity (RPM)	Torque (ft-lb)
55	74.40	1964	199
60	82.84	2143	203
65	91.70	2321	207
70	101.02	2500	212
75	110.83	2679	217
80	121.17	2857	223
85	132.06	3036	228

RPM: Revolution per minute

$$\text{Torque} = \frac{(\text{Horsepower})(550)}{\text{RPM}\left(\dfrac{2\pi}{60}\right)} \qquad (12.2)$$

Table 12.2 is the completed solution for the table. Since angular velocity and torque are known, the student would plot the points in Figure 12.1 and draw a line through them to complete the solution. The line goes through the most efficient portion of the engine map, so the thermal efficiency of the engine will be very good. If the average fuel consumption for this line is 200 g/kW-h, the average thermal efficiency is 69.5/200 = 34.75%. The line at 85 mph is just outside the engine map curve, so it would not be possible for this engine to produce 228 ft-lb of torque at 3036 RPM.

Example 12.2: If we assume a constant engine thermal efficiency of 34.75% during this acceleration, what would you expect the fuel economy to be during this acceleration? (138,700 BTU/gal for diesel fuel) [Note: Fuel economy during hard acceleration is not very good, even for a small diesel-powered car.] (**Table 12.3**)

Solution: The equations used to complete the table are (**Table 12.4**):

$$\text{Fuel Power} = \frac{\text{Total Horsepower}}{0.3475} \qquad (12.3)$$

TABLE 12.3 Example 12.2.

Speed (mph)	Total Horsepower	Fuel Power (hp)	Fuel Usage (gph)	Economy (mpg)
55	74.40			
60	82.84			
65	91.70			
70	101.02			
75	110.83			
80	121.17			
85	132.06			

TABLE 12.4 Completed Example 12.2.

Speed (mph)	Total Horsepower	Fuel Power (hp)	Fuel Usage (gph)	Economy (mpg)
55	74.40	214	3.93	14.00
60	82.84	238	4.37	13.72
65	91.70	264	4.84	13.43
70	101.02	291	5.33	13.13
75	110.83	319	5.85	12.82
80	121.17	349	6.40	12.51
85	132.06	380	6.97	12.19

© SAE International

$$\text{Fuel Usage} = \left(\text{Fuel Power}\right)\left(\frac{550}{778.17}\right)\left(\frac{3600}{138,700}\right) \qquad (12.4)$$

$$\text{Economy} = \frac{\text{Speed}}{\text{Fuel Usage}} \qquad (12.5)$$

Example 12.3: The car in question is traveling 70 mph when the transmission is in high gear and the engine is turning 2500 RPM. When cruising at constant speed the power requirements are much lower than when accelerating. Assume that the car is cruising at a constant speed with the horsepower requirements shown in the Table 12.5. The horsepower shown includes rolling and aero power and assumes that the transmission/drive system is 90% efficient in transmitting the power to the wheels. Plot the speed curve on the engine map in Figure 12.1 and estimate the thermal efficiency.

Solution: The torque numbers in the table are calculated using Eq. 12.2. From that, we locate the point on the graph and estimate the grams fuel required to produce 1

TABLE 12.5 Problem and solution for Example 12.3.

Speed (mph)	Total Horsepower	Angular Velocity (RPM)	Torque (ft-lb)	Efficiency (%)
40	7.13	1429	26.2	27.36
45	8.97	1607	29.3	27.45
50	11.14	1786	32.8	27.54
55	13.67	1964	36.5	27.64
60	16.58	2143	40.6	27.73
65	19.92	2321	45.1	27.82
70	23.72	2500	49.8	27.91

© SAE International

kW-h. At 40 mph the estimate used was 254 g/kW-h and at 70 mph it was estimated as 249 g/kW-h. The numbers in red in the table above are what was calculated for torque thermal efficiency of the engine. These are the answers students would generate on the exam. Notice that the thermal efficiency doesn't vary much over this range of speeds. I hope this helps the students understand why it is reasonable to assume the thermal efficiency is constant when driving on the highway.

Example 12.4: If the thermal efficiency is as shown in the table below, what would you expect the fuel economy to be during cruising at various speeds? (Diesel fuel has 138,700 BTU/gal)

Solution: The black numbers in **Table 12.6** are the problem statement and the red numbers are the solution. Eqs. 12.4, 12.5, and 12.6 are used to calculate the red numbers in the table. If we had used the average thermal efficiency the answers would be slightly different. Students usually ask if it is realistic for the car to get 83.71 mpg at 40 mph. The answer is yes. If you are willing to drive at a steady speed at 40 mph on a level track with the engine turning 1429 RPM in high gear the car will probably get very close to the 83.71 mpg in the table. The EPA highway estimate for this car is 44 mpg, which is approximately what the model yields at 70 mph.

TABLE 12.6 Problem and solution for Example 12.4.

Speed (mph)	Total Power (hp)	Efficiency (%)	Fuel Power (hp)	Fuel Usage (gph)	Economy (mpg)
40	7.13	27.36	26.05	0.478	83.71
45	8.97	27.45	32.68	0.600	75.06
50	11.14	27.54	40.45	0.742	67.39
55	13.67	27.64	49.45	0.907	60.62
60	16.58	27.73	59.81	1.097	54.68
65	19.92	27.82	71.62	1.314	49.47
70	23.72	27.91	85.00	1.559	44.89

© SAE International

Example 12.5: The typical acceleration when starting from a stoplight is 7 ft/s². If you are driving a 5400-lb truck, calculate the force (in pounds) for the 7 ft/s² acceleration.

Solution: Force is mass times acceleration:

$$\text{Force} = \left(\frac{5400\,\text{lb}}{32.2\,\text{ft/s}^2} \right) \left(7\,\text{ft/s}^2 \right) = 1174\,\text{lb} \qquad (12.6)$$

Example 12.6: Assume the 5400-lb truck is to accelerate from rest to 30 mph at a constant 7 ft/s². Fill in **Table 12.7** below using the units indicated. Consider only kinetic power. Ignore rolling and aero power.

Solution: The speed in mph is the only thing given in the table. Speed in ft/s is a unit conversion from the mph speed. The time column starts at zero, and since it is constant acceleration the number in the time column is the speed in ft/s divided by the 7 ft/s² acceleration. Power is the 1174-lb force multiplied by the speed in ft/s. The power is converted to watts. The energy used is the average power in watts multiplied by the time step, and it is totaled.

$$\text{Energy}_{i+1} = \text{Energy}_i + \left(\frac{\text{Power}_{i+1} + \text{Power}_i}{2}\right)\frac{\left(\text{Time}_{i+1} - \text{Time}_i\right)}{3600} \qquad (12.7)$$

The calculations are shown in red in the table. For this problem, we could have calculated the kinetic energy at each speed and that would be the energy required in column six. This works only because we are only considering kinetic energy in the problem. If we had included rolling resistance and aerodynamic losses this would not have worked, so Eq. 12.7 is presented as a more general way to calculate the energy in column 6.

TABLE 12.7 Problem and solution for Example 12.6.

Time (s)	Speed (mph)	Speed (ft/s)	Power (ft-lb/s)	Power (W)	Energy (W-h)
0	0	0	0	0	0
1.04762	5	7.333	8609	11,672	1.698
2.09524	10	14.667	17,217	23,344	6.793
3.14286	15	22	25,826	35,015	15.285
4.19047	20	29.333	34,435	46,687	27.173
5.23809	25	36.667	43,043	58,359	42.457
6.28571	30	44	51,652	70,031	61.138

Example 12.7: A mid-sized car has a weight of 4200 lb, Crr = 0.010, The frontal area is 1.68 m² and the drag coefficient is 0.33. The efficiency of the drive system on the car is 90%. A model is to be developed for the car accelerating from rest at 7 ft/s² going up a 3% grade. Fill in all of the empty boxes in the **Table 12.8**.

Solution: As with previous examples the numbers in black are given as part of the problem statement and you are to fill in the rest of the table. In column 3 the speed in ft/s is calculated as the speed at the beginning of the time step plus acceleration times the time step. The speed in column 4 is the speed in ft/s multiplied by (5280/3600).

Equations for rolling, aerodynamic, gravity, and kinetic watts are given previously in the book. For convenience, the equations are listed again below. W is in lb, V is in mph, and A is in m².

$$\text{Rolling Power} = (1.989)(W)(V)(Crr)\left(1 + \frac{V}{200}\right) \qquad (12.8)$$

TABLE 12.8 Problem and solution for Example 12.7.

Time (s)	Accel (ft/s²)	Speed (ft/s)	Speed (mph)	Rolling (W)	Aero (W)	Kinetic (W)	Gravity (W)	Total (W)
0	7	0	0	0	0	0	0	0
1	7	7	4.773	408.2	3.23	4343.5	1196.1	6612.3
2	7	14	9.545	835.5	25.83	13,030.5	2392.2	18,093.3

$$\text{Aero Power} = (0.05357)(V^3)(A)(Cd) \tag{12.9}$$

$$\text{Gravity Power} = (1.989)(W)(V)(\text{Grade}) \tag{12.10}$$

$$\text{Kinetic Power} = (0.0454)\frac{\left[V_{i+1}^2 + V_i^2\right]}{T} \tag{12.11}$$

The total power is obtained by adding the rolling, aero, kinetic, and gravity terms and dividing by the drive efficiency.

Example 12.8: As you move from college to your first job you are planning to drive your Ford Escape and pull a small U-Haul trailer behind it to move all of your worldly possessions. The Escape will be loaded to a weight of 4800 lb and the trailer will be loaded to 2500 lb. Assume that the rolling coefficient for the Escape and trailer is 0.01, the aerodynamic drag area is 2.8 m², and the drag coefficient is 0.65. The Escape can provide a maximum of 85 hp to the wheels. Calculate the steady-state speed for the Escape and trailer going up a 4% grade. Also, a 7% grade.

Solution: Converting, 85 hp is the same as 63,384.5 W. Since it is a steady-state speed the kinetic power will be zero. Gravity, rolling, and aerodynamic power can be calculated as a function of speed using Eqs. 12.8, 12.9, and 12.10. The sum is set equal to the 63,384.5 W power the Escape can provide to the wheels. Plugging in the parameters yields the following equation:

$$(145.2)(V)\left(1 + \frac{V}{200}\right) + (0.0957)(V^3) + (580.8)(V) = 63,384.5 \tag{12.12}$$

Solving Eq. 12.12 yields a steady-state speed of **57.9 mph** for the 4% grade. For a 7% grade, the only change is in the gravity term, and the steady-state speed is **45.4 mph**.

Example 12.9: For the ford escape of the previous example problem, detail out what happens for the first few time steps as the car comes to a 7% grade driving 75 mph. Fill in **Table 12.9** below.

   Solution: As with previous examples, the problem is the black text and the solution is in red. The initial speed is 75 mph. Gravity, rolling and aero W are calculated using Eq. 12.8, 12.9, and 12.10. The total will be more than the power the engine can provide at the wheels, so the difference must be made up of kinetic energy. Kinetic Joules is calculated as the energy difference divided by the time step of 0.5 s. V_2^2 is calculated using Eq. 11.31. This value is then used as the speed in the next row.

TABLE 12.9 Problem and solution for Example 12.9.

Time (s)	Speed (mph)	Gravity (W)	Rolling (W)	Aero (W)	Engine (W)	Kinetic (J)	$(V_2)^2$ (mph)^2
0	75	76,228	14,973	41,132	63,384.5	68,949	5521
0.5	74.30	75,520	14,797	39,996	63,384.5	66,927	5420
1.0	73.62	74,827	14,624	38,904	63,384.5	64,970	5322
1.5	72.95	74,147	14,456	37,853	63,384.5	63,072	5227

© SAE International

12.2 **Review of Units 1 and 2**

Example 12.10: We are developing an exercise workout, and part of the workout is to drag a box across the gym floor. The coefficient of friction between the box and the floor is 0.4. We want it to require about 100 W power to drag the box at 3 mph. How much should the box weigh? (**Figure 12.2**)

   Solution: The following equations are used:

$$\text{Drag Force} = (0.4)(W) \qquad (12.13)$$

$$\text{Speed} = (3\,\text{mph})\left(\frac{1609}{3600}\right) = 1.34\,\text{m/s} \qquad (12.14)$$

FIGURE 12.2 Person dragging a box across the gym floor.

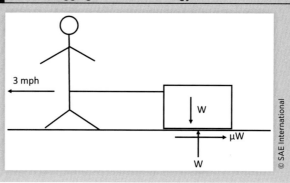

© SAE International

$$100 \, W = (0.4)(W)(1.34) \qquad (12.15)$$

The weight of the box should be 186.5 N, which is approximately **40 lb**. A person dragging a 40-lb box across the gym floor at 3 mph will require approximately 100 W of power.

Example 12.11: Assume 100 W output average during a 30-min workout. Muscles are approximately 20% efficient in converting the carbohydrate fuel into useful work. How many calories will be burned during the 30-min workout? (1 cal = 4184 J)

Solution: To produce 100 W with muscles that are 20% efficient requires (100/0.2) = 500 W of fuel consumption. The person will consume fuel at a rate of 500 W, which is 500 J/s. 30 min is 1800 s, so the total calorie consumption is calculated as:

$$\text{Calories Consumed} = (500 \, J/s)(1800 \, s)\left(\frac{\text{Cal}}{4184 \, J}\right) = 215 \, \text{cal} \qquad (12.16)$$

Working out at a rate of producing 100 W power for 30 min would be an intense workout for most people. The person would burn about 215 cal during the 30-min workout. This is the equivalent calories of a small candy bar, and less than the calories in a smoothie the person might purchase as a reward for the intense workout. Working out does not burn as many calories as we would like.

Example 12.12: A bicycle event is to be held where the riders will travel on the relatively level pavement for 50 miles. Three classes of riders will enter the event and the planners need to estimate how long it will take for the three classes to finish the race, so the finish line and banquet following the race can be planned. The three classes are:

   a. Wimpies – capable of providing 40 W power to the bicycle.
   b. Averages – capable of providing 90 W power to the bicycle.
   c. Athletes – capable of providing 180 W power to the bicycle.

The total drag force on the bicycle is given by the following equation:

$$\text{Drag Force} = (4 + (0.1)V^2) \, N, \text{ with V in mph} \qquad (12.17)$$

Estimate the time required for the three groups to finish the 50-mile event. We will need to have the finish line open and ready when the athletes arrive. The finish line will need to stay open until the wimpies have had time to finish so they can feel they are a part of the event, and the banquet will need to be planned late enough to allow everyone to get showered and ready.

Solution: Power is drag force multiplied by speed. The drag force equation comes out in N if the speed is in mph for Eq. 12.17. To get watts we need to multiply by the speed in m/s, so we need to use the unit conversion of (1609/3600). It is easy to miss that unit conversion in the problem. The average speed for the three classes of riders is calculated as:

a. $40\ W = (4 + 0.1\ V^2)\ V\ (1609/3600)$ $V = 8.26$ mph for Wimpies

b. $90\ W = (4 + 0.1\ V^2)\ V\ (1609/3600)$ $V = 11.57$ mph for Averages

c. $180\ W = (4 + 0.1\ V^2)\ V\ (1609/3600)$ $V = 15.07$ mph for Athletes

To travel 50 miles the wimpies will need 6.05 h, the averages will need 4.32 h and the athletes will need 3.32 h to complete the race. Assuming everyone starts the race at the same time, the finish line should open 3 h after the race starts and stay open for 4 h.

Example 12.13: The total drag force on a concrete canoe is 5 lb when traveling through the water at 6 mph. How much paddling power are the students providing in moving the canoe along at 5 mph? Express your answer in ft-lb/s, W, and hp.

$$\text{Power} = (5\,\text{lb})(6\,\text{mph})\left(\frac{5280}{3600}\right) = 44\ \text{ft-lb/s} \qquad (12.18)$$

Solution: The power required is 44 ft-lb/s. With unit conversions, this is equivalent to 0.08 hp or 60 W.

Example 12.14: Assume that the drag force on the canoe varies with the square of the speed of the canoe. How much power is required to push the canoe along at 2 mph? Express your answer in ft-lb/s, W, and hp.

Solution: Fluid drag varies with the square of speed in most cases. If the drag force is 5 lb at 6 mph and varies with the square of speed an equation can be developed for the drag force on the canoe at any speed:

$$\text{Drag Force} = (5\,\text{lb})\left(\frac{V}{6\,\text{mph}}\right)^2, \text{with V in mph} \qquad (12.19)$$

At 2 mph the drag force is 0.556 lb and the power is 1.63 ft-lb/s, 0.00296 hp, or 2.21 W. This is a very low power requirement. It would be easy to paddle the canoe along at 2 mph. A similar calculation can be done for 10 mph yielding 203 ft-lb/s or 276.2 W power required. When paddling the person does not really get to use his or her leg muscles and so the power a person can produce paddling is much less than on a bicycle or rowing with a sliding seat. It is difficult to paddle a canoe at 10 mph. Another way to look at the problem is to note that fluid drag power varies with the cube of speed. If the speed is reduced by a factor of 3 the power will be reduced by a factor of 27. The answers to this example problem are 1/27 of the answers to the previous problem.

Example 12.15: A small airplane cruises at 135 mph using 225 hp from the engine. The density of air is 1.2 kg/m³ at the altitude and temperature in question, and the propeller has a diameter of 1.9 m. Assume that the engine burns aviation fuel that has an energy density of 120,200 BTU/gal, and that the thermal efficiency of the engine is 28%.

a. Find the drag force on the airplane.

b. Find the fuel economy of the airplane in mpg.

Solution: To solve this problem we will need to use the propeller problems in Chapter 3, Eqs. 3.37, 3.38, 3.39, and 3.40. Substituting in the correct numbers yields the following:

$$V_1 = (135 \, \text{mph}) \left(\frac{1609}{3600} \right) = 60.34 \, \text{m/s} \tag{12.20}$$

$$V = \frac{60.34 + V_2}{2} \tag{12.21}$$

$$\text{Thrust Force} = \frac{1}{2}(1.2)\pi(0.95)^2 \left(V_2^2 - 60.34^2 \right) \tag{12.22}$$

$$\text{Power} = (\text{Thrust Force})(V) = (225)(745.7) \tag{12.23}$$

The solution is that V_2 = 71.659 m/s. This can be substituted into the thrust force equation to yield a drag force of 2542 lb.

To calculate fuel economy, we start by converting the 225 hp the engine produces into BTU/h, which is 572,528 BTU/h. The thermal efficiency of the engine is 28%, so it will need to burn fuel at a rate of:

$$\text{Fuel Consumption} = \left(\frac{572{,}528 \, \text{BTU/h}}{0.28} \right) \left(\frac{\text{gal}}{120{,}200 \, \text{BTU}} \right) = 17.01 \, \text{gph} \tag{12.24}$$

$$\text{Fuel Economy} = \frac{135 \, \text{mph}}{17.10 \, \text{gph}} = 7.94 \, \text{mpg} \tag{12.25}$$

Example 12.16: A jet airplane uses one gallon of fuel to go approximately 1.45 miles while flying 530 mph. Assuming that the engines are e = 35% efficient, calculate the drag force on the jet. Assume that the engine uses 70 kg of air for each kg of fuel. The energy density of the fuel is E_f = 46.6 MJ/kg. The fuel has a density of 2.88 kg per gal.

Solution: 530 mph is 236.88 m/s. The mass flow rate of the air is given to be 70 times the mass flow rate of the fuel. Letting m_a be mass flow rate of the air and m_f

be mass flow rate of the fuel, it follows that $m_a = 70\, m_f$. We calculate the velocity of the combustion products leaving the engine relative to the aircraft V_e from Eq. 4.26 as:

$$V_e = \sqrt{\frac{2(0.35)(46.6\times10^6)\dot{m}_f + 70m_f(236.88)^2}{70\dot{m}_f + \dot{m}_f}} = 717.5\,\text{m/s} \qquad (12.26)$$

We are given that the aircraft uses 1 gallon (2.88 kg) of fuel to go 1.45 miles. Since the speed is 530 mph we can calculate the mass flow rate of fuel for the aircraft. Once we know the mass flow rate of the fuel Eq. 4.23 is used to calculate the drag force on the jet.

$$m_f = \left(\frac{2.88\,\text{kg}}{1.45\,\text{mi}}\right)(530\,\text{mph})\left(\frac{h}{3600\,\text{s}}\right) = 0.424\,\text{kg/s} \qquad (12.27)$$

$$\text{Drag} = \big[0.424 + 70(0.424)\big](717.5) - 70(0.424)(236.88) = 14{,}569\,\text{N} \qquad (12.28)$$

The thrust of the engine is 14,569 N, and for steady-state flight this must equal the drag force on the airplane.

Example 12.17: According to the Department of Energy, we used 28.3 Quads of energy in transportation in 2018. Because of the thermal efficiency, about 22.4 Quads are rejected to the atmosphere as heat, and 5.9 Quads are converted into useful mechanical energy to power the vehicles. The EPA estimates that transportation generates 1800 million metric tons of carbon dioxide emissions each year. Based on this information, calculate the number of grams of carbon dioxide generated for each kW-h of useful mechanical energy in transportation.

Solution: First convert the 5.9 Quads into kW-h energy.

$$(5.9\times10^{15}\,\text{BTU})\left(\frac{\text{kW-h}}{3412\,\text{BTU}}\right) = 1.729\times10^{12}\,\text{kW-h} \qquad (12.29)$$

1800 million metric tons is 1830×10^{12} grams of carbon dioxide. Dividing the numbers:

$$\frac{1800\times10^{12}\,\text{g CO}_2}{1.729\times10^{12}\,\text{kW-h}} = 1041\,\text{g CO}_2 \text{ per kW-h} \qquad (12.30)$$

In Unit 2, we calculated that on average electric power plants generate 458 g CO_2 per kW-h of electricity. The purpose of this problem was to illustrate that electric power plants are much more efficient than the engines used in transportation for the amount of carbon dioxide generated.

Example 12.18: Assume that the atmosphere has a mass of 5.15×10^{18} kg and that the average molecular weight is 29. How many moles of gas are in the atmosphere?

$$\left(5.15 \times 10^{18} \text{ kg}\right)\left(\frac{1000 \text{ g}}{\text{kg}}\right)\left(\frac{\text{mole}}{29 \text{ g}}\right) = 1.776 \times 10^{20} \text{ moles} \qquad (12.31)$$

Example 12.19: Assume that the world produces 3.29×10^{13} kg of carbon dioxide annually and that the molecular weight of carbon dioxide is 44. How many moles of carbon dioxide are put into the atmosphere each year?

$$\left(3.29 \times 10^{13} \text{ kg}\right)\left(\frac{1000 \text{ g}}{\text{kg}}\right)\left(\frac{\text{mole}}{29 \text{ g}}\right) = 1.134 \times 10^{15} \text{ moles} \qquad (12.32)$$

Example 12.20: Based on the two numbers above, how many parts per million does the carbon dioxide represent in the atmosphere? [Data indicate that the carbon dioxide content of the atmosphere is increasing at about 2.2 ppm annually.]

$$\frac{1.134 \times 10^{15}}{1.776 \times 10^{20}} = 6.385 \times 10^{-6} = 6.385 \text{ ppm} \qquad (12.33)$$

Carbon emissions from burning fossil fuel each year is enough to be 6.385 ppm in the atmosphere. Measurements indicate the carbon dioxide concentration to be increasing at 2.2 ppm. Where is the rest of it going? The oceans absorb carbon dioxide and plants use carbon dioxide in photosynthesis.

Example 12.21: We are using a liquid diesel fuel that on average has 24 hydrogen atoms for each 12 carbon atoms in the chemical composition. Draw a suitable molecule for $C_{12}H_{24}$.

Solution: There are many possibilities. One possibility is shown in **Figure 12.3**.

FIGURE 12.3 $C_{12}H_{24}$ molecule.

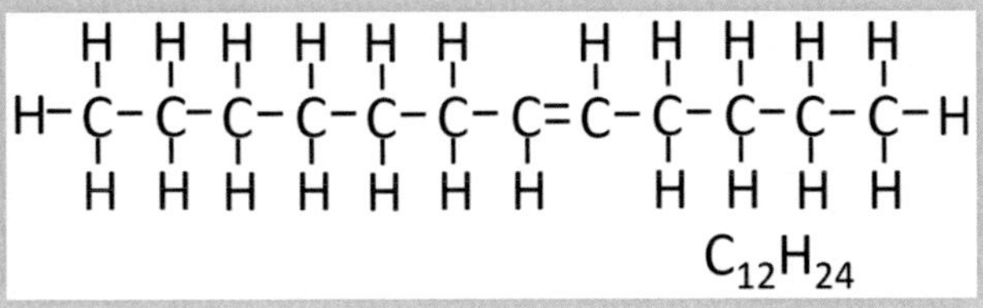

Example 12.22: The density of the $C_{12}H_{24}$ fuel is 812 g per liter and the energy density is 135,000 BTU per gal. Write the chemical reaction for the combustion of the fuel and calculate the number of BTUs heat energy we get from the fuel for each gram of carbon dioxide generated.

$$C_{12}H_{24} + 18\,O_2 \rightarrow 12\,CO_2 + 12\,H_2O \tag{12.34}$$

The chemical balance equation tells us that one mole of fuel will produce 12 moles of carbon dioxide (and also 12 moles of water). The molecular weights of the fuel and the 12 carbon dioxide atoms are calculated as:

$$C_{12}H_{24} = 12(12) + 24(1) = 168\,g \tag{12.35}$$

$$12\,CO_2 = 12\left[12 + 2(16)\right] = 528\,g \tag{12.36}$$

The conclusion is that 168 g of fuel will produce 528 g of carbon dioxide. Knowing that the heat value of the fuel is 135,000 BTU/gal and the density is 812 g/liter it is possible to calculate the number of BTU heat energy per gram of fuel.

$$\left(135,000\,BTU/gal\right)\left(\frac{0.26417\,gal}{liter}\right)\left(\frac{liter}{812\,g\,fuel}\right) = 43.92\,BTU/g\,fuel \tag{12.37}$$

And since 168 g fuel produces 528 g carbon dioxide:

$$\left(\frac{43.92\,BTU}{g\,fuel}\right)\left(\frac{168\,g\,fuel}{528\,g\,CO_2}\right) = 13.97\,BTU/g\,CO_2 \tag{12.38}$$

Example 12.23: A company produces a small wind turbine for home use. The turbine blade is 8 ft in diameter and the efficiency of the turbine is estimated to be 45%. Assume that the density of air is 1.2 kg/m³. The generator is 90% efficient in converting the energy from the turbine into electric energy. How much power do you estimate the turbine will produce in a 15-mph wind?

Solution: First calculate V_1, the speed of the wind as 6.704 m/s. The radius of the turbine blade is 4 ft, which is converted to 1.219 m. Since efficiency is known to be 45%, the efficiency formula (Eq. 7.10) can be used to calculate V_2.

$$0.45 = \frac{1}{2}\left[1 - \frac{V_2^2}{6.704^2}\right]\left(1 + \frac{V_2}{6.704}\right) \tag{12.39}$$

Solving the equation yields $V_2 = 4.571$ m/s. The average velocity V is 5.6375 m/s. The turbine power is equal to:

$$\text{Turbine Power} = \frac{1}{2}(1.2)(5.6375)\pi\left(1.219^2\right)\left(6.704^2 - 4.571^2\right) = 379.75\,W \tag{12.40}$$

The generator is 90% efficient in converting the turbine power into electricity. The power delivered in a 15-mph wind is 341.8 W.

Example 12.24: Based on a value of $0.13 per kW-h, how long would it take for the wind turbine to produce $1.00 worth of electricity if it operates in a 15-mph wind?

Solution: Assuming the wind blows steady at 15 mph it takes $(1000/341.8) = 2.93$ h to produce 1 kW-h of electric energy. At $0.13 per kW-h it takes 7.69 kW-h to be worth $1.00. The answer is that it takes 22.5 h to produce $1.00 worth of electricity.

Putting up the tower and wind turbine and getting the electricity wired would be expensive. At a rate of producing $1.00 worth of electricity every 22.5 h it will produce $3,983.00 worth of electricity in 10 years. To break even, the homeowner could spend about this amount in installing the wind turbine and getting it hooked up.

index